Azospirillum III

Genetics · Physiology · Ecology

Proceedings of the
Third Bayreuth Azospirillum Workshop

Edited by Walter Klingmüller

With 102 Figures

Springer-Verlag
Berlin Heidelberg New York Tokyo

Prof. Dr. Walter Klingmüller
Lehrstuhl für Genetik
Universität Bayreuth
Universitätsstraße 30
8580 Bayreuth, FRG

The workshop was sponsored by
BASF Ludwigshafen
Universitätsverein Bayreuth
Bayerisches Staatsministerium für Landwirtschaft und Forsten

ISBN-13:978-3-642-70793-3 e-ISBN-13:978-3-642-70791-9
DOI: 10.1007/978-3-642-70791-9

Library of Congress Cataloging in Publication Data. Main entry under title: Azospirillum III. Held on June 20–21, 1985 at the University of Bayreuth, West Germany. Bibliography: p. Includes index. 1. Azospirillum – Congresses. I. Klingmüller, W., 1929-. II. Title: Azospirillum three. QR82.A9A97 1984 589.9'5 85-25009.

2131/3130-543210

PREFACE

On June 20th and 21st 1985 the third workshop on "Azospirillum: Genetics, Physiology, Ecology" took place at the University of Bayreuth, West Germany, organized by the genetics department. There were about 80 participants, who came from German research institutions, from other European countries, from India, Egypt, North and South America. The former workshops had taken place in Bayreuth too in 1981 and 1983 respectively, hence the organizers could draw on the experiences then obtained.

Azospirilla have, during the past 12 years, found an ever increasing scientific interest, because first, these soil bacteria carry the genetic information for binding molecular nitrogen from the air, and second, they live in close vicinity to the roots of grain crops and forage grasses. By exploitation of these two properties, it is hoped to develop inoculation procedures in soils poor in nitrogen.

The reports on the first afternoon focussed, as a result of the Bayreuth research interest, on genetic approaches. Considerable progress has been made, via the application of refined gene technological methods, in the understanding of nitrogen fixation. Restriction maps for the structural nif-genes in both *Azospirillum brasilense* and *Azospirilum lipoferum* were presented. Also cloning of genes for histidin-biosynthesis and glutamine synthetase was reported. Broadly discussed - on both days - were further metabolic properties, and those features of Azospirilla that seem important for affecting or colonizing the roots of the host plants, e.g. growth hormon production, preferred utilization of carbon sources, denitrification, chemotaxis.

On the following morning, a number of reports brought data, supporting the contention that inoculation of plants, not only in the greenhouse, but also in field trials, can indeed result in growth responses, e.g. increased root development, shoot height, and yield in general. However, these reports up to now only bear on limited regional soil and climatic conditions. To obtain responses, a suboptimal supply with mineral nitrogen fertilizer seems crucial. An obvious trend was, to work with better controlled and faster in vitro systems instead of field inoculations. Still, early effects thus to be registered on seedlings do not guarantee beneficial effects on adult plants in the field. More work on this point is therefore urgently needed.

During the workshop, an important conceptual breakthrough, documented by several sets of data, was an older idea that responses upon inoculation may depend not only on the Azospirillum strain, but also to a large degree on the plant cultivar studied, or even on the individual genotype of single plants. The term "nitrogen fixation supporting", viz. nis-phenotype, will no doubt spread to plant geneticists very soon and tempt them into cooperation with Azospirillum workers on this subject.

The number of contributions, as well as the number of participants of the workshop, had increased considerably, compared to the former workshops. This documents the need for such workshops and the importance of the subject. The same can be concluded from the increased number of observers sent by chemical and fertilizer companies. The opportunity for an intense exchange of experiences and ideas was exploited not only during the scientific meetings, but also during informal events. Again, many new personal contacts were established. It is hoped that these will pay off in the future. The foreign participants who had come for the first time were impressed by Bayreuth University, its new research institutions, and the genetics department. An evening at the summer residence "Eremitage" and its park added positively to the overall impression.

Thanks to the speakers, who had been asked to bring the manuscripts with them ready for print, and thanks to the publisher, it is again possible to bring these proceedings on the market within three months after the workshop. They will help the participants to recapitulate and better understand the contributions. They should moreover help all other interested people, to judge the present state of Azospirillum work, and to evaluate the prospects of working with this organism in the future.

W. KLINGMÜLLER

CONTENTS

Construction of a Gene Bank of Azospirillum brasilense
R. FANI, M. BAZZICALUPO, E. GALLORI, L. TURBANTI and M. POLSINELLI (With 2 Figures) 1

The nifHDK Operon of Azospirillum brasilense Sp7
B. PERROUD, S.K. BANDHARI and C. ELMERICH (With 2 Figures). 10

Problems and Prospects of Site-Directed Transposon Mutagenesis in Azospirillum
M. SINGH and W. KLINGMÜLLER (With 3 Figures) 20

Cosmid Cloning of Nitrogenase Structural Genes of Azospirillum lipoferum
R. FAHSOLD, M. SINGH and W. KLINGMÜLLER (With 4 Figures) .. 30

Construction of a Genomic Library of Azospirillum brasilense Sp7 and Cloning of the Glutamine Synthetase Gene
C. FOGHER, H. BOZOUKLIAN, S.K. BANDHARI and C. ELMERICH (With 4 Figures) 41

Cloning of the Gene for the Restriction Enzyme AbrI from Azospirillum brasilense ATCC 29711
G. SCHWABE and W. KLINGMÜLLER (With 4 Figures) 52

Stability of Broad Host Range Plasmids in Azospirillum Cloning of a 5.9 kBp Plasmid from A. brasilense RO7
K. MICHIELS, M. VANSTOCKEM, J. VANDERLEYDEN and A. VAN GOOL (With 4 Figures) 63

Transfer and Random Integration of Tn5 in Azospirillum
M. VANSTOCKEM, K. MICHIELS, J. VANDERLEYDEN and A. VAN GOOL (With 5 Figures) 74

Mutation in Azospirillum brasilense
M.M. DEL GALLO, L. GRATANI and G. MORPURGO (With 6 Figures) 85

Production of Indol-3-Acetic Acid by Azospirillum brasilense
S. HOREMANS and K. VLASSAK (With 4 Figures) 98

Auxin Production by Azospirillum brasilense Under Different Cultural Conditions
R. DE FRANCESCO, G. ZANETTI, P. BARBIERI and E. GALLI (With 3 Figures) 109

Comparison of Nitrogenase Regulation in A. brasilense, A. lipoferum and A. amazonense
A. HARTMANN, H.-A. FU, S.-D. SONG and R.H. BURRIS (With 5 Figures) 116

Some Physiological and Biochemical Properties of Denitrification by Azospirillum brasilense
G. DANNEBERG, W. ZIMMER and H. BOTHE (With 2 Figures) 127

An "in vitro" Assay for Evaluating the Azospirillum Wheat Association
M. BAZZICALUPO, E. CRESTA and F. FAVILLI (With 2 Figures). 139

Secretion and Cleavage of Sucrose by Wheat: A System of Chemotactic Attraction and Nitrogenase Induction in Azospirillum lipoferum
D. HESS, D. HEINRICH, J. KEMMNER and S. JEKEL (With 11 Figures) 147

Field Inoculation of Rice Using in vitro Selected Bacterial and Plant Genotypes
P.B.B.N. CHARYULU, F. FOURCASSIE, A.K. BARBOUCHE, L. RONDRO HARISOA, A.M.N. OMAR, P. WEINHARD, R. MARIE and J. BALANDREAU (With 3 Figures) 163

Evaluation of Nitrogen Fixation in Sorghum Cultivars Inoculated with Different Strains of Azospirillum spp.
C. CHRISTIANSEN-WENIGER, R.M. BODDEY and J. DÖBEREINER (With 1 Figure) 180

Possible Contribution of Azospirillum spp. to the Nutritional Status of Wheat Plants Grown in Sandy Soils of Gassim-Saudi Arabia
N.A. HEGAZI and H. SALEH (With 7 Figures) 189

Root Colonization and Plant Growth of Grasses and Cereals in Unsterile Soil and Nitrogenase (C_2H_2) Activity of Sterile Cereal Seedlings after Inoculation with Azospirillum spp.
G. JAGNOW (With 2 Figures) 203

Response of Plant Roots to Inoculation with Azospirillum brasilense and to Application of Indole Acetic Acid
W. KOLB and P. MARTIN (With 3 Figures) 215

Electron Microscopy of Young Wheat Roots Inoculated with Azospirillum
J.H. WHALLON, G. ACKER and H. EL-KHAWAS (With 10 Figures). 222

Cysts of Azospirilla under Various Cultural Conditions
L. SADASIVAN and C.A. NEYRA (With 15 Figures)............. 230

Pleomorphism in Azospirillum
J.H. BECKING 243

ERRATUM (Balandreau Azospirillum Workshop II, 1983) 263

LIST OF PARTICIPANTS

Abd-el-Salam, M.	Cairo	Egypt
Acker, G.	Bayreuth	Germany
Alef, K.	Bayreuth	Germany
Balandreau, J.	Vandoeuvre	France
Bazzicalupo, M.	Firenze	Italy
Beck, Th.	München	Germany
Becking, J.H.	Wageningen	Netherlands
Beyse, J.	Bayreuth	Germany
Bogdahn, U.	Bayreuth	Germany
Bothe, H.	Köln	Germany
Bozouklian, H.	Paris	France
Charyulu, P.	Andhra Pradesh	India
Christiansen-Weniger, C.	Rio de Janairo	Brasil
Danneberg, G.	Köln	Germany
De Coninck, K.	Leuven	Belgium
Del Gallo, M.	Rome	Italy
Dreessen, R.	Leuven	Belgium
El-Khawas, H.	Cairo	Egypt
Fages, J.	Toulouse	France
Fahsold, R.	Bayreuth	Germany
Fani, R.	Firenze	Italy
Favilli, F.	Firenze	Italy
Fendrik, I.	Hannover	Germany
Fogher, C.	Paris	France
Galli, E.	Milano	Italy
Gasparrini, G.	Rome	Italy
Grossmann, K.	Limburgerhof	Germany
Hartmann, A.	Madison	USA
Hegazi, N.A.	Cairo	Egypt

Hess, D.	Stuttgart	Germany
Hoffmann, R.	Bayreuth	Germany
Horemans, S.	Leuven	Belgium
Jagnow, G.	Braunschweig	Germany
Kleiner, D.	Bayreuth	Germany
Klingmüller, W.	Bayreuth	Germany
Kolb, W.	Stuttgart	Germany
Kraft, I.	Bayreuth	Germany
Kröckel, L.	Bayreuth	Germany
Krumm, M.	Stuttgart	Germany
Laine, M.	Vandoeuvre	France
Markus, P.	Bonn	Germany
Marocco, A.	Piacenza	Italy
Martin, P.	Stuttgart	Germany
Michiels, K.	Leuven	Belgium
Morpurgo, G.	Rome	Italy
Neuray, J.	Ougree	Belgium
Niemann, E.G.	Hannover	Germany
Ong, E.	Leuven	Belgium
Perroud, B.	Paris	France
Powell, K.	Jealott's Hill	England
Rahmen, M.	Dhaka	Bangladesh
Ripke, H.M.	Erlangen	Germany
Schwabe, G.	Bayreuth	Germany
Singh, M.	Bayreuth	Germany
Stolp, H.	Bayreuth	Germany
Vanderleyden, J.	Leuven	Belgium
Vanstockem, M.	Leuven	Belgium
Whallon, J.	East Lansing	USA
Zanetti, G.	Milano	Italy
Zimmer, W.	Köln	Germany

CONSTRUCTION OF A GENE BANK OF AZOSPIRILLUM BRASILENSE

R. Fani, M. Bazzicalupo, E. Gallori, L. Turbanti and M. Polsinelli
Dipartimento di Biologia Animale e Genetica
University of Florence, Via Romana 17, Florence, Italy

Summary: A gene bank of DNA from *A.brasilense*, strain Sp6, has been constructed in the single EcoRI and BglII sites of plasmid pRK290. The vector was able to be transferred by conjugation from *E.coli* to *A.brasilense* by using the helper plasmid pRK2013 at a frequency of about 10^{-3} per recipient cell. Clones carrying the *his* and *pyr* genes have been identified by genetic complementation; the expression of *A.brasilense* in *E.coli* has been demonstrated. Keywords: gene bank, cloning, conjugation.

Introduction

The nitrogen fixing bacterium *A.brasilense* stimulated in the recent years considerable interest because of its ability to fix nitrogen in association with the roots of economically important cereals (1, 2).
Genetic studies have been also carried out in order a)to map genes, b) to manipulate the genome and c) to study the molecular basis of the regulation of nitrogen fixation (3, 4, 5, 6).
The only way to introduce genetic material in *A.brasilense*, described so far, is conjugation mediated by P group incompatibility plasmids (3,5). For this reason such kind of conjugative plasmids seems the better choice as vectors for molecular cloning in *A.brasilense*. One of this plasmids is pRK290 (7), a small

Azospirillum III: Genetics, Physiology, Ecology
Ed. by W. Klingmüller

derivative of RK2,that has been succesfully used for the construction of a genomic bank in Rhizobium meliloti and has also been used as carrier to introduce nif genes in Azospirillum(8). In the present report we describe the construction of a genomic Azospirillum DNA bank in the vector pRK290 and the use of genetic complementation for cloning genes. The experiments reported also demonstrate that Azospirillum genes can be expressed in E.coli.

Materials and Methods

Bacterial strains and plasmids are listed in Table 1

Media. A.brasilense strains were grown either in minimal medium MSP (9) or in rich medium (Antibiotic Medium n.3 Difco). E.coli minimal medium was MMD (14), while rich medium was LB (12); medium for transformation selection was LM (10).

DNA isolation and manipulation. A.brasilense Sp6 chromosomal DNA was purified from 200 ml rich medium culture as described (7); large scale pRK290 DNA preparation from E.coli DH1 was as described (11) followed by Ethidium Bromide Cesium Chloride gradient.

Chromosomal DNA was partially digested with either EcoRI or BglII and size fractionated on sucrose gradient. 10-20 Kb fragments were ligated to pRK290 DNA completely digested with the same restriction endonuclease and dephosphorilated with calf intestinal alkaline phosphatase. Ligated DNA was then used to transform DH1 competent cells as described (10). Plasmid DNA from 1.5 ml cultures of E.coli and A.brasilense was also extracted by using the rapid alkaline lysis method (11).

Conjugation. Conjugations were carried out on filter membrane as described (15) using plasmid pRK2013 in E.coli DH1 as helper.

Table 1. Bacterial strains and plasmids

Bacteria/plasmids	Relevant genotype	
A.brasilense		
Sp6	prototroph	(9)
SPF13	*his rif*	Sp6 + EMS
SPF69	*his ade*	"
SPF12	*leu*	"
E.coli		
DH1	*rec*A1 *thi*-1	(10)
HB101	*rec*A13 *pro*A2	(11)
CSH54	*pyr*F *his thi trp*	(12)
FB1	Δ *his*750 *gnd*	C.B.Bruni
FB190	*his*G2743 *rec*A56	"
FB251	*his*B855 *rec*A56	"
FB181	*his*I903	"
FB182	*his*F892	"
FB184	*his*A915	"
Plasmids		
pRK290	Tc^r	(7)
pRK2013	Km^r	(13)

Results

Construction and characterization of the bank. A genomic bank of DNA from *A.brasilense* strain Sp6 has been constructed in the vector plasmid pRK290 as described in Methods. *Azospirillum* DNA fragments were inserted both in EcoRI and BglII unique restriction sites. Plasmids from tetracycline resistant transformants were extracted by rapid method and cleaved with EcoRI or BglII in order to measure on gel electrophoresis the average size of inserts. Results of one of these experiments are reported in Figure 1; the analysis of 80 plasmids revealed that about 80 per cent of them carried inserts with a mean size of about 10 kb, ranging from 6 to 15 kb. Some E.coli DH1 carrying recombinant plasmids were grown for more than 20 generations in the absence of antibiotic and the DNA was then analized as described; results showed that the plasmids were quite stable and that the Azospirillum DNA inserts were always mantained whithout modifications. Conjugation experiments using pRK290 and its recombinant derivatives as donors, pRK2013 as helper and A.brasilense as recipient, were carried out looking for the optimal conditions for high frequency of plasmid transfer. Results obtained indicated that a) there was no difference in the conjugation frequency of pRK290 and derivatives with Azospirillum DNA inserts, b) the optimal temperature was 30°C, and c) the frequency of transfer raises during the first 10-12 hours of the mating up to about 5 x 10^{-3} per recipient cell (Figure 2).

Cloning genes of Azospirillum. DNA of Azospirillum carried by the vector pRK290 can be analized with classic genetic techniques, by conjugation in A.brasilense and by transformation and conjugation in E.coli, supposing that Azospirillum genes can express in E.coli.

In order to test the possibility to use the described bank to identify genes, we have chosen 200 indipendent transformant clones, subdividing them in four pools and using them as donors in

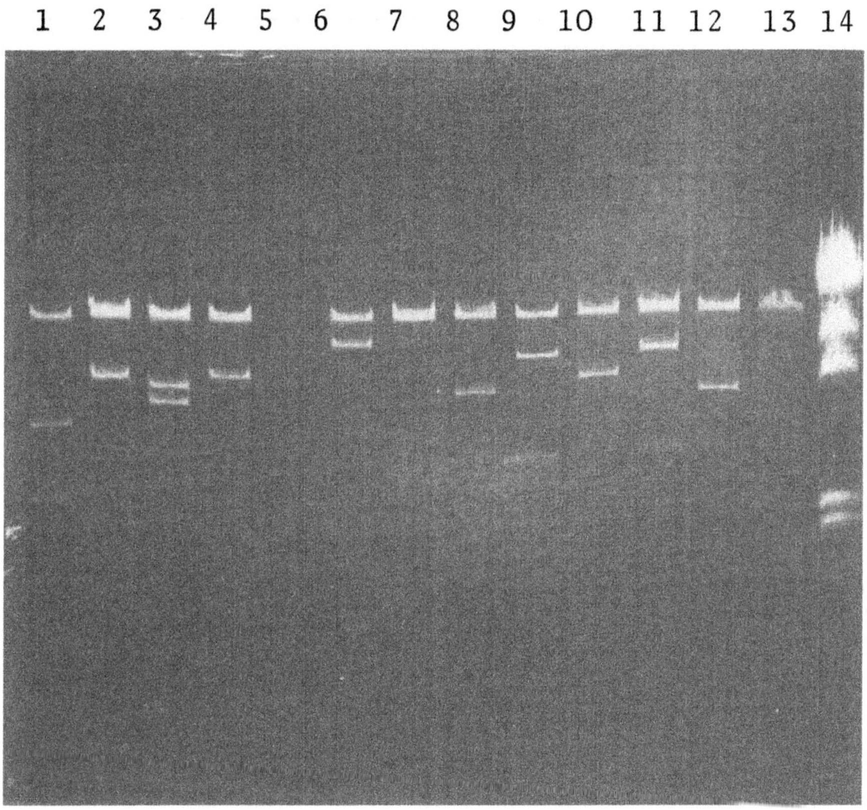

Figure 1. EcoRI-digested plasmid DNAs from the *A.brasilense* gene bank. The uppermost band in each lane is pRK290 (lanes 1-12). Lane 13 represents pRK290 digested with EcoRI and lane 14 is a total digest of lambda DNA with HindIII.

conjugation experiments with *Azospirillum* strains SPF12 *leu* and SPF13 *his* as recipients. Results of the experiments, reported in Table 1, indicated that the pool n.2 included one or more clones carrying histidine gene. In the following experiments we divided the second pool in five subpools which were conjugated again with strain SPF13; the subpool complementing

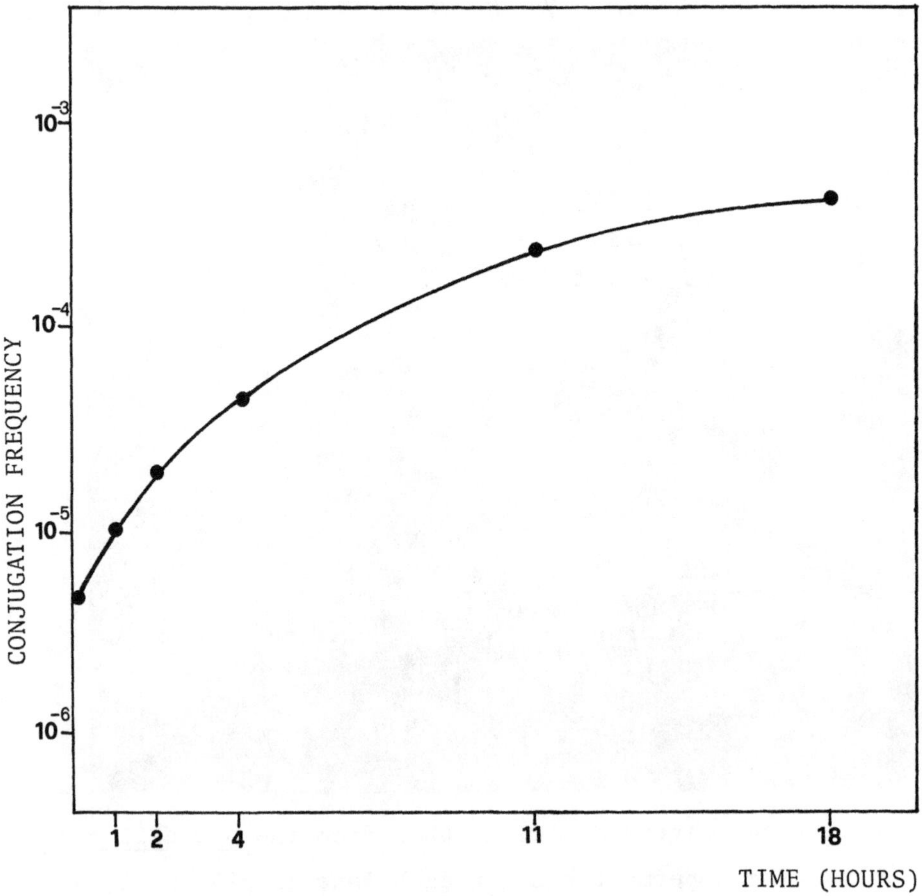

<u>Figure 2.</u> Effect of mating lenght on the frequency of conjugation between *E.coli* DH1 (pRK290) and *A.brasilense* Sp6.

his mutation was further subdivided and single clones mated again until we found a clone carrying a plasmid called pAF58. This plasmid contained the *his* gene in a DNA fragment of about 9 kb inserted in the EcoRI site of pRK290.

The presence of *Azospirillum his* gene in the plasmid pAF58 was further confirmed by transforming *E.coli* HB101 with pAF58 DNA and co-

Table 2. Conjugal transfer of 200 recombinant plasmids of the bank from *E.coli* DH1 to *A.brasilense* mutants

Donor	Recipient	Selected marker(s) (colonies/plate)		
		Tc^r	Tc^rLeu^+	Tc^rHis^+
DH1 (pool 1)	SPF 12 *leu*	350	0	-
+ DH1 (pRK2013)	SPF 13 *his*	1320	-	0
DH1 (pool 2)	SPF 12 *leu*	300	0	-
+ DH1 (pRK2013)	SPF 13 *his*	1800	-	20
DH1 (pool 3)	SPF 12 *leu*	800	0	-
+ DH1 (pRK2013)	SPF 13 *his*	2640	-	0
DH1 (pool 4)	SPF 12 *leu*	1540	0	-
+ DH1 (pRK2013)	SPF 13 *his*	2300	-	0

Tc, tetracycline; leu, leucine, his, histidine.
Each pool contained 50 different clones.

njugating again in SPF13 as shown in Table 2. Furthermore pAF58 also complemented another independent *his* mutation of *A.brasilense* strain SPF69.

Azospirillum genes express in E.coli. Plasmid pAF58 was further used to study the expression of *Azospirillum* genes in *E.coli*. Plasmid DNA was extracted and used to transform six *E.coli*

Table 3. Conjugal transfer of pAF58 from E.coli HB101 to A.brasilense SPF 13

Donor	Recipient	Selected marker(s) (colonies/plate)	
		Tc^r	Tc^rHis^+
HB101 (pAF58) + DH1 (pRK2013)	SPF 13 his	150	145

Tc, tetracycline; his, histidine

strains carrying his mutation in five different genes, hisG, hisB, hisI, hisF and hisA and a complete his operon deletion. Results obtained showed that plasmid pAF58 complemented hisB mutation of E.coli indicating that the Azospirillum DNA insert included at least the gene(s) coding for the enzymes Imidazole glycerol phosphate dehydrase and Histidinol phosphate phosphatase. Acridine curing experiments on E.coli FB251 hisB(pAF58) demonstrated that His^+ phenotype was 100% linked to tetracycline resistance.

The results demonstrated that at least some genes of Azospirillum can be expressed in E.coli. Since genes of Azospirillum express in E.coli we tried to use transformation to identify genes of the bank. Plasmid DNA extracted from 200 clones divided into 10 pools was used to transform E.coli strain CSH54; tetracycline resistant transformants were tested for complementation of Trp, Ura, Thi mutations: two out of 50 transformants from one of the pools showed to be Ura$^+$. Also in this case the Ura$^+$ phenotype was completely linked to tetracycline resistance.

References

1. Dobereiner J. (1983). In W.Klingmuller (ed.)pp. 9-23. Azospirillum II. Exs 48; Birkhauser, Basel.
2. Okon J. (1985). In: N_2 fixation and CO_2 methabolism. P.W. Ludden and S.B.Burno eds. pp. 165-174. Elsevier Science Pub.Corp.Inc.
3. Polsinelli,M., Baldanzi,E., Bazzicalupo, M. and Gallori E. (1980). Molec.G.Genet. 178: 709-711
4. Quiviger, B.,Franche,C.,Lutfalla,G.,Rice,D.,Haselkorn,R. and Elmerich C.(1982). Biochimie 64: 495-503
5. Franche,C.,Canelo,E.,Gauthier,D. and Elmerich C.(1981). FEMS Microbiol.Lett. 10: 199-202
6. Pedrosa,F.D. and Yeates M.G.(1984). FEMS Microbiol.Lett. 23: 95-101
7. Ditta,G.,Stanfield,S.,Corbin,D. and Helinski D.R.(1980). P.N.A.S. 77: 7347-7351
8. Nair,S.K.,Jara,P.,Quiviger,B. and Elmerich C. (1983). In W.Klingmuller (ed.) pp.29-38.Azospirillum II. Exs48; Birkhauser, Basel.
9. Bani,D.,Barberio,C.,Bazzicalupo,M.,Favilli,F.,Gallori,E. and Polsinelli M. (1980). J.Gen.Microbiol.119: 239-244
10. Hanahan,D. (1983).J.Mol.Biol. 166: 557-580
11. Maniatis,T., Fritsch,E.F. and Sambrook J. (1982).Molecular cloning. CSH 1982.
12. Miller J. (1972). Experiments in molecular genetics.CSH
13. Figurski,D. and Helinski,D.R. (1979) . P.N.A.S. 76: 1648-1652
14. Davis, B.D. and Mingioli E.S. (1950).J.Bacteriol. 60:17-28
15. Ries,G., Holloway B.W. and Puhler A. (1980) Genet.Res. 36: 99-109

THE NIFHDK OPERON OF AZOSPIRILLUM BRASILENSE Sp7

B. Perroud, S.K. Bandhari and C. Elmerich

Unité de Physiologie Cellulaire, Département de Biochimie et Génétique Moléculaire, Institut Pasteur, 28 rue du Docteur Roux 75724 Paris cedex 15, France

Summary : The transcriptional organization of the *nifHDK* cluster of *Azospirillum* was examined by transposon Tn*5* mutagenesis, genetic complementation, and product analysis. Tn*5* mutagenesis was performed in *E. coli* on plasmid pAB3, a derivative of pSUP202 in which the *nifHDK* region was cloned. This plasmid can be transferred but cannot replicate in *Azospirillum*. Twenty six insertions located in different positions of the *nifHDK* region were obtained and recombined in the *Azospirillum* genome. Insertions in *nifH*, *nifD* and *nifK* led to a Nif^- phenotype. Our results are in agreement with the existence of a single operon transcribed in the *nifHDK* order as in *Klebsiella pneumoniae*. Key words : *nifHDK* cluster, Tn*5* mutagenesis, *nif* products, transcriptional organization.

Introduction

In *Klebsiella pneumoniae*, a cluster of 17 nitrogen fixation genes organized in 7 or 8 transcription units has been identified. The nitrogenase structural genes, *nifHDK*, are carried on a 6.2 kb *Eco*RI fragment and the 3 genes are transcribed in a single operon (for a review, see 1). Using homology with a *K. pneumoniae* *nifHDK* probe, a 6.7 kb *Eco*RI fragment, designated AbRI, was cloned from total DNA of *A. brasilense* Sp7 (2). Heteroduplex analysis performed with the *nifHDK* cluster of *K. pneumoniae*, established the approximate location of the corresponding *Azospirillum* *nifH,-D,-K* genes on this fragment (2). The AbRI fragment was subcloned in the broad host range vector pRK290 (3), and after construction of partial diploids in *Azospirillum*, complementation of a few Nif^- mutants was observed (4). In order to establish the transcriptional organization of the *Azospirillum* *nifHDK* cluster, site directed mutagenesis was performed using transposon Tn*5*. The methodology developed by Simon *et al* (5) in *Rhizobium meloliti* was applied to *Azospirillum*. These authors constructed

Azospirillum III: Genetics, Physiology, Ecology
Ed. by W. Klingmüller

plasmid pSUP202, a derivative of pBR325 that contained the site for mobilization (mob region) of P type plasmids. Plasmid pSUP202 cannot replicate outside the enteric bacteria, and can be used as a suicide vehicle to introduce specific markers in a large number of Gram-negative bacteria (5).

Materials and Methods

Strains and plasmids

Azospirillum strain was *A. brasilense* Sp7 (ATCC29145) (6). Other strains were *Escherichia coli* K12 derivatives. Plasmids were kept in C600 (*trp*, *leu*, *lacY*, *recA*, *rif*, r^-m^-) or in JC5466 (*trp*, *his*, *recA*, *rpsE*, *rif*). Strain S17.1 (*pro*, *thi*, RP4 (Tc::Mu Km::Tn*7*) contains an RP4 derivative inserted in the chromosome, which is no more transferable but which can complement in trans Tra^- plasmids containing the "Mob" site (such as pSUP202) (5). The derivative of S17.1 containing Tn*5* in the chromosome, referred to as S17.1::Tn*5*, was constructed by P. Mazodier. Plasmids are listed in Table 1. The AbRI fragment is the 6.7 kb *Eco*RI fragment cloned from Sp7, and which carries the *nifHDK* region (2).

TABLE 1 Plasmid phenotypes

Plasmid	Vector	Phenotype	Reference
pRK2013		Km^R Tra^+	(3)
pSUP202	pBR325	Amp^R, Cm^R, Tc^R, Mob	(5)
pRK290		Tc^R, IncP, Tra^-	(3)
pAB1	pACYC184	Tc^R, AbRI	(2)
pAB3	pSUP202	Amp^R, Tc^R, Mob, AbRI	(this work)
pAB35	pRK290	Tc^R, AbRI	(4)
pAB36	pAB35	Tc^R, 2.5 kb *Pst*I deletion of the *nifDK* genes	(4)
pPC1201	PGV822	Amp^R, 0.9 kb *Eco*RI-*Kpn*I *K. pneumoniae* *nifH* fragment	(7)

Media, growth conditions and assay for nitrogen fixation.

Complete media were nutrient broth for Azospirillum and Luria broth for E. coli (4). Minimal and N-free media for Azospirillum were made of the salt basis "K" (8). The carbon source was sodium lactate 5 g/l, and the nitrogen source was NH_4Cl 20 mM. Two assays were used to determine nitrogenase activity in whole cells. One corresponded to the assay under conditions of nitrogenase derepression (9). The other was performed as follows. Two drops of culture were inoculated in a 10 ml "bijou" bottle containing 7 ml of N-free medium supplemented with 1 mg ml^{-1} glutamate. Glutamate does not repress nitrogen fixation in Azospirillum (10). The bottles were capped and incubated for 16 h at 30°C with shaking. Then 1 ml acetylene was added and ethylene production was measured after 1 h.

Isolation of Tn5 insertions in the host genome.

Plasmid construction, transformation, DNA isolation and restriction analysis employed conventional techniques (11). Plasmid pAB3 was obtained after cloning the AbRI fragment at the EcoRI site of pSUP202. Plasmid pAB3 was introduced into S17.1::Tn5 by transformation and several independent transformants were crossed with JC5466. Transconjugants were selected on LB medium supplemented with kanamycin (Km: 20 μg/ml), tetracycline (Tc: 10 μg/ml) and rifampicin (Rif: 50 μg/ml). The frequency of Km^RTc^R among Km^STc^R clones was 10^{-5}. The Tn5 insertions were localized in pAB3, after extraction of the plasmids by the method of Birnboim and Doly (12), and restriction analysis. Plasmids carrying Tn5 in AbRI were subsequently introduced into the strain Sp7, using pRK2013 as mobilizing agent in a triparental cross (3,4). Azospirillum recombinants were selected on minimal medium containing Km (20 μg/ml). Co-integration of pAB3 in the host genome was tested by assaying Tc resistance. The percentage of Tc^RKm^R clones ranged from 5 to 50% and varied with the location of Tn5 in AbRI.

Hybridization

DNA hybridization was performed on Southern blots with ^{32}P nick-translated probes using conventional techniques (11).

Pulse labeling of cells proteins

Azospirillum cells were incubated under conditions of nitrogenase derepression, and were labelled after 3 h with 5 μC of ^{14}C aminoacid mixture for 20 min. Cells were collected and protein extracts were separated on SDS-PAGE as described (13). After impregnation for fluorography, gels were exposed to XRay films.

Results and discussion

Physical mapping of *nifH*

By heteroduplex analysis the approximate location of *nifH*, was found in the 2.8 kb *Eco*RI-*Pst*I fragment of AbRI. A more precise location using a *nifH* specific probe established that part of the *nifH* sequence was located in the 0.7kb *Xho*I-*Pst*I fragment, as shown in Figures 1 and 2a. This experiment was performed by probing pAB1 restriction fragments with an internal part of the *K. pneumoniae* *nifH* gene. The probe used corresponded to codons 71 to 206 (14), purified as a 400 bp *Ava*II-*Kpn*I fragment from plasmid pPC1201.

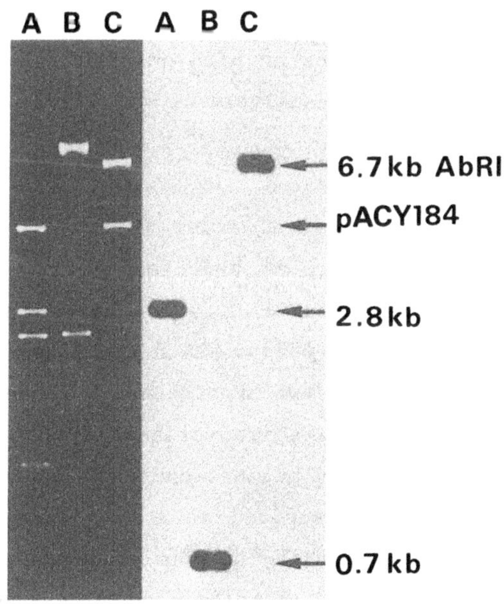

Figure 1 . Physical mapping of *nifH* in the AbRI fragment. Left part : pAB1 hydrolysis by *Pst*I-*Eco*RI (A); *Xho*I-*Pst*I (B); *Eco*RI (C); right part corresponding autoradiogram after hybridization with the *K. pneumoniae* *nifH* *Ava*II-*Kpn*I fragment probe.

Phenotypes of Tn*5* insertions in the *nifHDK* region.

Over 300 Tn*5* insertions were isolated in the pAB3 plasmid. Figure 2a shows the localization of 26 different Tn*5* insertions in the AbRI fragment. These insertions were recombined in the Sp7 genome. Nitrogen fixation expressed in percentage of the value obtained with the wild type is reported in Figure 2b. Tn*5* insertions leading to a Nif^+ phenotype were localized at each extremity of the AbRI fragment. In addition two others were isolated close to the *Pst*I site. Insertions elsewhere led to a Nif^- phenotype. For example 4 insertions, 27, 9, 62, and 66 were located in the *nifH* coding sequence. The size of the fragment between insertions 27 and 66 is 0.8 kb, and is compatible with a polypeptide product of 35kd. A similar comment can be done for insertions located in the *nifD* and *nifK* region. Hybridization with a Tn*5* probe was performed with total DNA extracted from a few Nif^- and Nif^+ strains. Results were in agreement with the localization found in pAB3 (data not shown). In particular, this was done for insertion 13 which led to a Nif^+ phenotype, and which is located between *nifH* and *nifD*. This latter observation suggested that the *nifHDK* cluster might be composed of two transcription units. However, since reinitiation from Tn*5* promoter was demonstrated in a few cases (15), the polarity effect of Tn*5* insertions on *nifH* on *nifD* and *nifK* was determined by genetic complementation and, analysis of *nif* polypeptide synthesis.

Complementation analysis

Partial diploids were obtained after introduction of plasmid pAB35 and pAB36 into the Km^R mutants. As expected pAB35, which contained the entire *nifHDK* cluster, complemented all the Nif^- mutants (see Table 2). Concerning complementation with pAB36, which contained an intact *nifH* gene but which is deleted for *nifDK*, two hypotheses could be envisaged.

i) The *nifH* and *nifDK* genes are transcribed separately, and pAB36 will complement the *nifH* mutants but not the *nifD* and *nifK* mutants.

ii) The *nifHDK* genes are transcribed as a single operon and pAB36 will not complement any of the *nifH*, *nifD* or *nifK* mutants.

As reported in Table 2, no complementation was observed with pAB36 and this result is in favor of the second hypothesis.

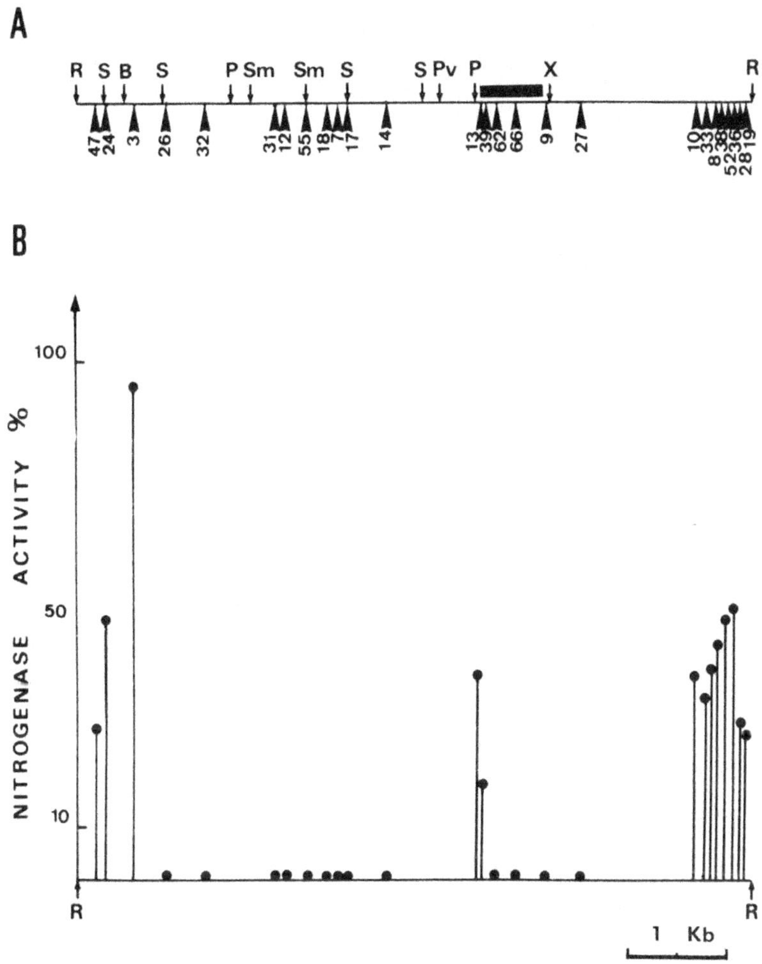

Figure 2. Localization and Nif phenotype of the Tn5 insertions.
A: Physical mapping of the insertions in the AbRI fragment; restriction sites: B: BglII; P: PstI; Pv: PvuII; R: EcoRI; S: SalI; Sm: SmaI; X: XhoI; black triangles indicate the position of the Tn5 insertions; ▬ indicate the region of homology with *K. pneumoniae nifH*.
B: Nitrogenase activity of the Km^R mutants obtained after recombination of the insertions in the genome of *Azospirillum* Sp7; nitrogenase activity was determined by the "bijou" bottle assay method, results are expressed in % of the wild type. 100% nitrogenase activity correspond to 60 nmol/min/mg protein.

TABLE 2 Complementation analysis

Insertion number	Nif Phenotype in Azospirillum	Nitrogenase activity % no plasmid	pAB35	pAB36
no insertion	+	100	72	70
19	+	28	70	30
10	+	39	75	40
27	-	<1	66	<1
9	-	<1	73	<1
39	+	18	70	ND
13	+	39	72	ND
14	-	<1	50	ND
17	-	<1	40	<1
07	-	<1	70	<1
18	-	<1	70	<1
12	-	<1	60	<1
31	-	<1	62	<1
32	-	<1	50	<1
26	-	<1	45	<1
03	+	95	80	82
24	+	50	55	45

Nitrogenase activity was assayed under derepression conditions; 100 % activity correspond to 60 nmol/min/mg protein; ND: not determined.

Product analysis

The molecular weight of Azospirillum nitrogenase subunits was previously established by immunoprecipitation with immunsera prepared against the MoFe-protein and the Fe-protein of K. pneumoniae (16). Azospirillum nitrogenase MoFe-protein subunits were estimated at 60 and 64 kd. The Fe-protein, which is likely composed of a single subunit gave two bands of 39 and 36kd. As Azospirillum is subject to ammonia switch-off (9; P. Jara unpublished results), the 39 kd product was tentatively assigned to the covalently modified Fe-protein (17).

Analysis of products made by Tn5 insertions mutants was in agreement with the previous identification of the nitrogenase polypeptides, though the 39 kd polypeptide was never observed. A typical autoradiogram is shown in Fig. 3. Insertions in nifH abolished the synthesis of the 36, 60 and 64 kd products showing that insertions in nifH were polar on nifD and nifK. The 36 kd product was detected in all the nifD and nifK mutants, and corresponded to the nifH product. In nifD mutant the 60 and 64 kd products disappeared and it was expected to find one of these two polypeptides in the nifK mutants. Unfortunately, the 60 and 64kd polypeptides were not detected in the nifK mutants (Figure 3). This may be due to the instability of the 60 kd polypeptide. The same situation was encountered in K. pneumoniae (13) where it was suggested that the nifD product was instable in the absence of the nifK product. Though some ambiguity remains, it is likely that nifD corresponds to the 60 kd product and that nifK corresponds to the 64 kd one.

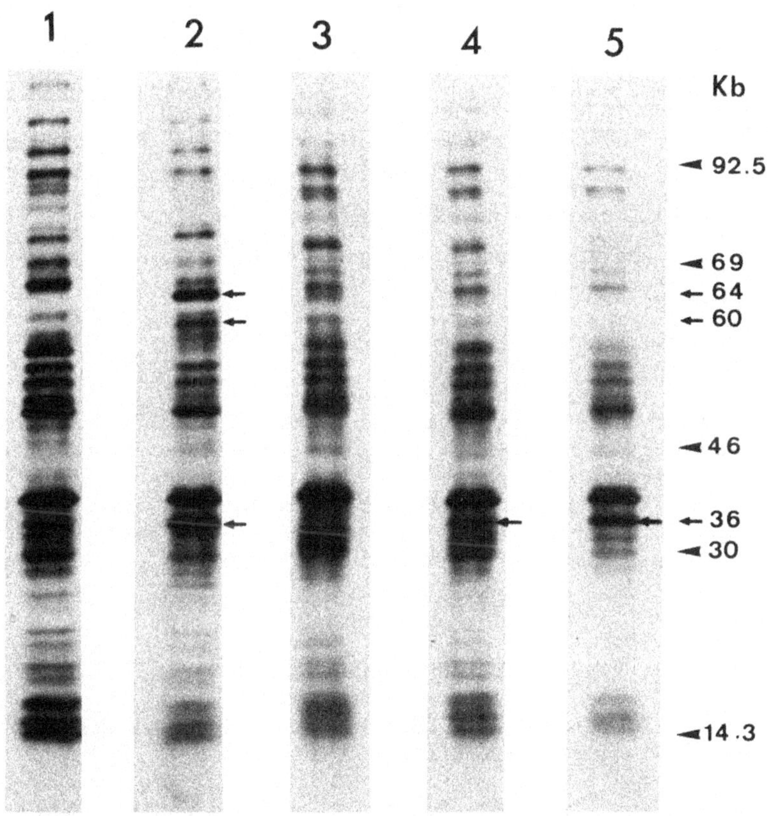

Figure 3. Analysis of the polypeptides synthesized by Sp7 and Tn5 insertions mutants. Cells samples labelled with C aminoacid were collected and protein extracts were analyzed by SDS-PAGE (5-15% gradient polyacryamide), 1 and 2: Sp7; 3 to 5: insertion mutant 9, 14, 26; 1: labeling under conditions of ammonia assimilation; 2 to 5: labeling under conditions of nitrogenase biosynthesis.

To understand why insertions 13 and 39 gave a Nif^+ phenotype the mutants were labelled in the presence and in the absence of ammonia. If one assume that in insertions 13 and 39 nif transcription proceeds from a Tn5 promoter or from the creation of a promoter-like sequence due to the Tn5 insertion (see Ref. 15), the transcription of the nifDK polypeptides should escape the ammonia control, whereas nifH is still repressed by ammonia. This was indeed the case. Insertion 13 mutant strain produced the 36, 60 and 64 kd polypeptide, in the absence of ammonia, and only the 60 and 64 kd polypeptides in the presence of ammonia.

In conclusion, genetic complementation and product analysis were in agreement with the transcription of the nifHDK genes as a single operon. Apart K. pneumoniae, organization of the nitrogenase structural genes in one operon was also reported in other diazotrophs e.g. Rhodopseudomonas capsulata (18), Azotobacter (19), and Rhizobium meliloti (20). This is not however a general rule, since in other cases, e.g. Anabaena (21) and R. japonicum (22) the nifHDK genes are not contiguous.

Acknowledgements

B.P. was the recipient of a predoctoral fellowship from the Ministère de l'Industrie et de la Recherche, S.K.B. was a postdoctoral fellows from the French Government.

This work was supported by funds from the University Paris VII, and by a research contract from Elf Bio-Recherche, Entreprise Minière et Chimique, Rhône-Poulenc Recherche, and CDF Chimie.

References

1. Elmerich, C. (1984). BIO/TECHNOLOGY 2, 967-978

2. Quiviger, B., Franche, C., Lutfalla, G., Rice, D., Haselkorn, R. and Elmerich, C. (1982). Biochimie 64, 495-502

3. Ditta, G., Stanfield, S., Corbin, D. and Helinski, D. (1980). Proc. Natl. Acad. Sci. USA 77, 7347-7351

4. Jara, P., Quiviger, B., Laurent, P. and Elmerich, C. (1983). Can. J. Microbiol. 29, 968-972

5. Simon, R., Priefer, . and Pühler, A. (1984). BIO/TECHNOLOGY 1, 784-789

6. Tarrand, J.J., Krieg, N.R. and Döbereiner, J. (1978). Can. J. Microbiol. 24, 967-980

7. Sibold, L., Pariot, D., Bhatnagar, L. and Aubert, J.P. (1985). Mol. Gen. Genet. in press

8. Franche, C. and Elmerich, C. (1981). Ann. Microbiol. (Inst. Pasteur) 132A, 3-19

9. Gauthier,D. and Elmerich, C. (1977). FEMS Microbiol. Lett. 2, 101-104

10. Pedrosa, F. and Yates, M.G. (1984). FEMS Microbiol. Lett. 23, 95-101

11. Maniatis T. Fritsch E.F., Sambrook J. (1982) Molecular cloning, Cold Spring Harbor Laboratory

12. Birnboim, H.C. and Doly, J. (1979). Nuc. Acids Res. 7, 1513-1523

13. Houmard, J., Bogusz, D., Bigault, R. and Elmerich, C. (1980). Biochimie, 62, 267-275

14. Scott, K.F., Rolfe, B. and Shine, J. (1981). J. Molec. Appl. Genet. 1, 71-81.

15. Berg, D.E., Weiss, A., and Cross land, L. (1980). J. Bacteriol. 142, 2, 439-446.

16. Nair, S. K., Jara, P., Quiviger, B. and Elmerich, C. (1983). In Azospirillum II, W. Klingmuller ed. Experentia Supplementum EXS 48, 29-38

17. Ludden, P.W. and Burris, R.H. (1978). Biochem. J. 175, 251-259

18. Avtges, P., Scolnik, P.A. and Haselkorn, R. (1983). J. Bacteriol. 156, 251-256

19. Jones, R., Woodley, P. and Robson, R. (1984). Mol. Gen. Genet. 197, 318-327

20. Szeto, W.W., Zimmerman, J.L., Sundaresan, V. and Ausubel, F.M. (1984). Cell 36, 1035-1043

21. Rice, D., Mazur, B. and Haselkorn, R. (1983). J. Biol. Chem. 257, 13157-13163

22. Kaluza, K., Fuhrmann, M., Regensburger, B. and Hennecke, H. (1983). J. Bacteriol. 155, 915-918

PROBLEMS AND PROSPECTS OF SITE-DIRECTED TRANSPOSON MUTAGENESIS IN AZOSPIRILLUM

Mahavir Singh and Walter Klingmüller
Lehrstuhl für Genetik, Universität Bayreuth
Universitätsstraße 30, D-8580 Bayreuth, FRG

Summary: Different vectors and transposons were tested to obtain site-directed nif⁻ mutants of *A. brasilense*. Tn5 proved more suitable than Tn1725 in this respect. A novel method of site-directed nif-gene deletion formation in combination with resistance-gene-cartridge mutagenesis is also reported.
Keywords: nif-genes, transposon mutagenesis, cloning, cartridge mutagenesis, nif⁻ mutants.

Introduction

Azospirillum is a widely distributed soil bacterium living in close proximity to plant roots where it is assumed to play an important role by virtue of its capacity to fix nitrogen and to produce plant growth substances (1, 2). Molecular analysis of nitrogen fixation (nif) genes and other related genes in Azospirillum is progressing steadily (3, 4, 5, 6, 7). The recently developed technique of site-directed mutagenesis is an invaluable tool for studying the organisation and function of genes in Azospirillum especially because as yet no method is available for transformation or transduction in this bacterium. Cloning, transposon-mutagenesis in *E. coli* and back transfer of the mutagenized genes into the wild type to produce predetermined mutations are the essential features of site-directed mutagenesis or reverse genetics (8, 9, 10).

There are a number of plasmid vector-systems available for site-directed mutagenesis in Gram-negative bacteria distantly related to *E. coli*. These vectors are essentially of two types: 1. Vectors replicating stably, both in the donor and in the

Azospirillum III: Genetics, Physiology, Ecology
Ed. by W. Klingmüller

recipient bacterium, e.g. pRK290, and 2. vectors unable to replicate in the recipient bacterium, e.g. pACYC184, pBR322. In the first case, the mutagenized genes are transferred to the wild type strain and the stably replicating vector is removed by using another plasmid belonging to the same incompatibility group, thus making the detection of the mutagenized phenotype easier. The second case is an ingenious way to avoid the problem of using the second plasmid, as the vector of mutagenized genes is lost automatically after transfer into the recipient.

Since no information exists about suitable vectors and transposons for site-directed mutagenesis in Azospirillum, we have tried two different replicons and two transposons. Of course, the main objective of the present study was to produce site-directed nif⁻ mutants. We report here a very successful application of cartridge-mutagenesis in order to produce well defined stable nif⁻ mutants in Azospirillum.

Materials and methods

Bacterial strains and plasmids: Bacterial strains and plasmids used in the present study are listed in Table 1.

Media and growth conditions: Complete medium for *E. coli* and Azospirillum was LB (12); minimal medium for Azospirillum (MAZ) was as described by Albrecht and Okon (13). *In vivo* nitrogenase activity was measured by C_2H_2 reduction in nitrogen-free, semi-solid MAZ (0,2 % agar). The following concentrations of antibiotics were used: Chloramphenicol (Cm), 25 µg/ml; Kanamycin (Kan), 40 µg/ml; Tetracycline (Tc), 15 µg/ml; Rifampicin (rif), 100 µg/ml.

Eppendorf-conjugation: We developed a simple and rapid method of conjugating bacteria in an 1.5 ml Eppendorf tube. Overnight cultures of donor and recipient were mixed and pelleted in a 'Hettich' table centrifuge. After washing the pellet twice with

Table 1. Bacterial strains and plasmids

Bacteria	Relevant characteristics	Source/ reference
E. coli		
HB101	Sm^r, r^-, m^-, $recA^-$	J. Collins, FRG.
S 17-1	Pro^-, r^-, m^+, RP4-2-Tc::Mu, Km::Tn7, Tp^rSm^r	(9)
MSO-8	Pro^-, r^-, m^+, RP4-2-Tc::Mu, Km::Tn7, Tp^rSM^r, Tn1725	This lab.
Azospirillum brasilense		
29145 rif	a rif^r derivative of ATCC 29145 wild type	This lab.
29710 rif	a rif^r derivative of ATCC 29710 wild type	This lab.
Plasmids		
pACYC 184	Tc^r, Cm^r	(11)
pSUP202	Tc^r, Cm^r, Amp^r, Mob^+	(9)
pAB1	Tc^r, Abr nifHDK	(3)
pMS175-2	Tc^r, Amp^r, Mob^+, Abr nifHDK	This study
pMS205	Tc^r, Amp^r, Mob^+, Abr nifHΔDK	This study
pMS188	Tc^r, Amp^r, Mob^+, Abr nif188::Tn5	This study

LB, the bacteria were suspended in 20-40 µl LB, centrifuged for 2 min. at 4000 rpm and incubated overnight at desired temperature (usually 30°) in the same Eppendorf tube. Next day the conjugation mixture was taken up in 1 ml LB, vortexed and plated on selective plates. Transconjugants were usually visible after 2-3 days.

DNA biochemistry: Isolation, purification and other manipulations of DNA were as described earlier (13, 14, 15).

Results

pACYC184 as vector and Tn1725 as transposon:

We previously reported Tn1725 mutagenesis of *Azospirillum brasilense nif* genes (Abr *nif*HDK) cloned in pAB1 (12). The exact locations of Tn1725 insertions were localized by comparing EcoRI and EcoRI + BglII digestions of the pAB1::Tn1725 plasmids. Since there is one asymmetrically located BglII site on the 6.7 kb EcoRI fragment of pAB1, the EcoRI + BglII double digestion compared to EcoRI digestion (there are two EcoRI sites present on the left and right ends of Tn1725) helped in mapping of the transposon insertions of Tn1725 (Fig. 1). We transferred the mutagenized plasmids (pAB1::Tn1725) by conjugation (mobilisation with pRK2013) into *A. brasilense* wild type strain 29145 rif selecting for chloramphenicol resistance. 40 colonies from each transfer were tested for nitrogen fixation (acetylene reduction). None was defective in nitrogen fixation. About 90% of the chloramphenicol (Cm^r) resistant colonies were also resistant to tetracycline (Tc^r) indicating that plasmid pAB1::Tn1725 had integrated in the chromosome (cointegrate formation). Thus, no nif^- mutant could be obtained. The other (10 %) Cm^r Tc^s transconjugants were most probably due to transposition of Tn1725 to new locations in the Azospirillum chromosome.

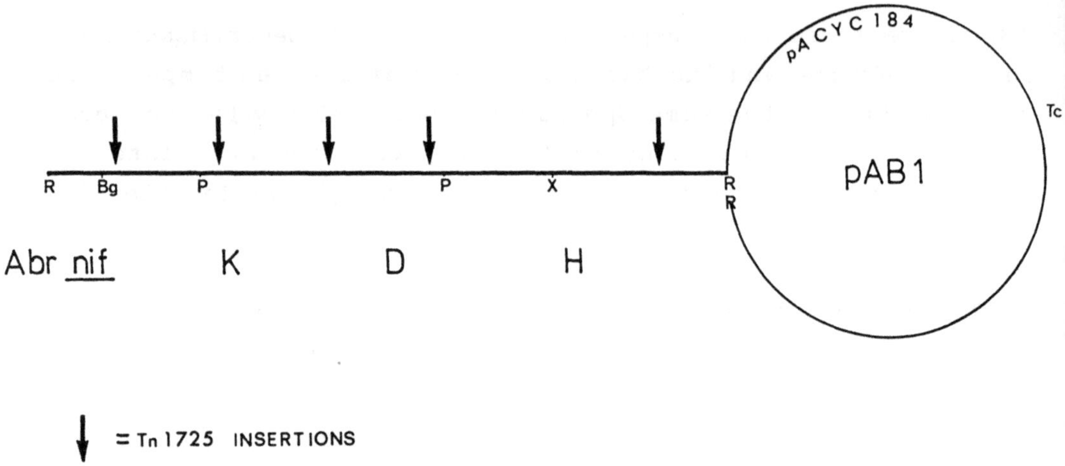

Figure 1. Location of Tn1725 insertions in the 6.7 kb EcoRI(Abr*nif*HDK) fragment of pAB1. The vector in pAB1 is not drawn to scale. Bg=BglII; P=PstI; R=EcoRI; X=XhoI.

pSUP202 as vector and Tn1725 as transposon:

Since pAB1 did not yield any nif$^-$ transconjugants in *A. brasilense*, we decided to try the pBR325-based mobilisable vector pSUP202 (9). For this, we cloned the 6.7 kb EcoRI fragment (Abr *nif*HDK) from pAB1 in pSUP202 and named the recombinant plasmid pMS175-2. The Abr *nif* genes on pMS175-2 were then mutagenized with Tn1725 directly in MSO-8, a derivative of the mobilizing strain S17-1 (details to be published elsewhere). We obtained insertions of Tn1725 at different sites on Abr *nif*HDK fragments; all insertions were located about 1.0 kb away from the left and the right end of the Abr *nif* fragment. The exact locations of Tn1725 insertions were mapped (Fig. 2) as described in the previous section.

The pMS175-2::Tn1725 derivatives were transferred to *A. brasilense* wild type strain 29145 rif. The insertion pMS175-2::Tn 1725 No. 8 (in *nif*H) yielded nif$^-$ transconjugants. That the nif$^-$ phenotype was due to site-directed mutagenesis of wild type *nif*HDK was confirmed by Southern hybridization between the chromosomal DNA of one nif$^-$ transconjugant and the radioactively

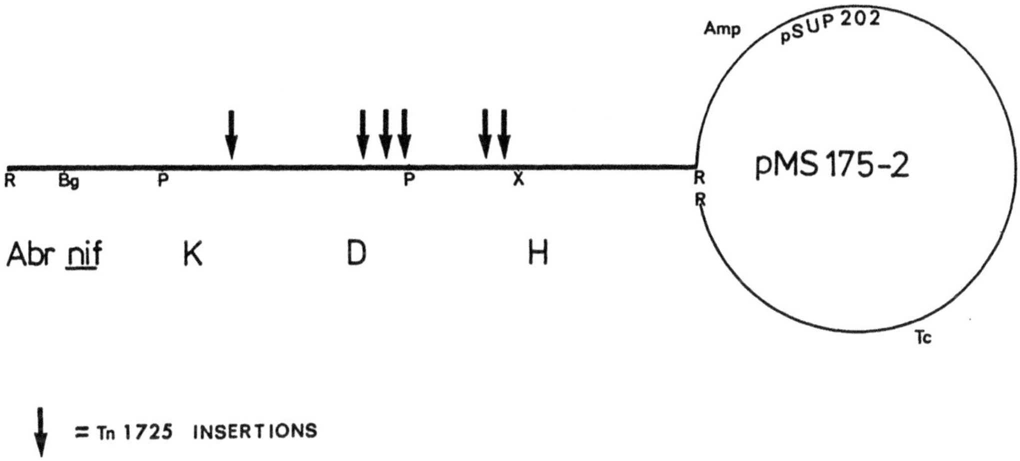

Figure 2. Location of Tn1725 insertions in the 6.7 kb EcoRI insert (Abr*nif*HDK) of pMS175-2. The vector pSUP202 is not drawn to scale. Bg=BglII; P=PstI; R=EcoRI; X=XhoI.

labelled Abr *nif*HDK fragment isolated from pAB1. All the Cm^r transconjugants were Tc^S showing that unlike the case of pAB1, no co-integrate formation between pMS175-2 derivatives and the *A. brasilense* chromosome had taken place. The high proportion of nif^+, Cm^r transconjugants indicates high transposition rates of Tn1725 in Azospirillum and a low rate of double crossing-over between the Abr *nif*HDK::Tn1725 insertions on pMS175-2 and the Abr *nif*HDK present in the chromosome of the wild type.

pSUP202 as vector and Tn5 as transposon

We have recently been successful in carrying out transposon (Tn5) mutagenesis in *A. brasilense* (details to be published elsewhere). A 23 kb EcoRI fragment from the genome of one of the Tn5 induced nif^- mutants was cloned in pSUP202 (Fig. 3). The recombinant plasmid pMS188 thus contains as yet unidentified nif-genes. We used pMS188 for site-directed mutagenesis in the wild type 29710 rif. Following transfer of pMS188 into 29710 rif, Kan^r transconjugants were obtained, purified and tested for acetylene

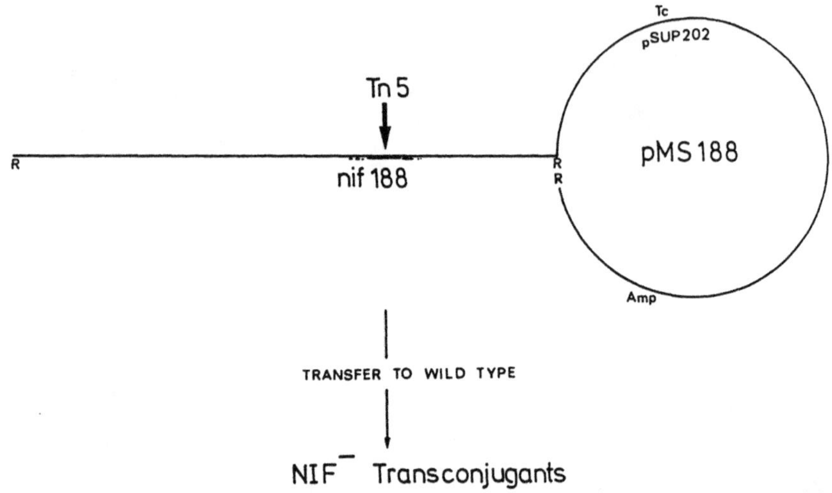

Figure 3. Cloning of a 23 kb EcoRI fragment containing Tn5 (from nif⁻ mutant no. 188). The vector pSUP202 is not drawn to scale. R=EcoRI.

reduction. A high percentage (about 50 %) of Kan^r transconjugants proved to be nif⁻. All the Kan^r transconjugants were Tc^s indicating that no co-integrate formation had taken place. These results further confirmed the linkage between the nif⁻ phenotype in mutant 188 and the insertion of Tn5.

Resistance-gene-cartridge mutagenesis:

This method involves cloning of a DNA fragment containing a gene for antibiotic resistance into the genes to be mutagenized (16). The resistance-gene-cartridge is unable to transpose and hence no interference occurs due to further transposition in the genome of the recipient.

The 6.7 kb EcoRI fragment in pMS175-2 (Fig. 2) was digested partially with Pst1 and ligated to a 4.0 kb Pst1 fragment containing Kan^r gene of Tn903. In one of the recombinant plasmids (pMS205) the 2.4 kb PstI fragment containing Abr *nif*DK is replaced by the 4.0 kb Kan^r fragment of Tn903. When we used the plasmid pMS205 for site-directed mutagenesis in *A. brasilense* wild type strain 29710 rif, all the 20 Kan^r transconjugants tested proved completely defective in acetylene reduction (0.01 % to 0.001 % activity of the wild type). Since these mutants contain a deletion of Abr nifDK, no reversion to nif$^+$ phenotype is expected to take place. Also, since the Kan^r gene is unable to transpose, no secondary nif$^-$ or other types of mutants should arise due to Kan^r-cartridge.

Discussion

A comparison of different vectors and transposons for site-directed mutagenesis in Azospirillum is presented. The plasmid pACYC184, although mobilizable by pRK2013 to Azospirillum, did not yield any nif$^-$ transconjugants but predominantly gave rise to co-integrates which even after removing the antibiotic selection did not result in nif$^-$ phenotype. Using a pSUP202 and Tn1725 combination we were more successful in getting the mutant transconjugants. It seems that the large size or some other feature of Tn1725 interferes with double crossing over which is essential for producing the nif$^-$ phenotype in the present series of experiments. It can be argued that some of the Tn1725 insertions (in pAB1 or pMS175-2) tried were in intergeneric regions or that there could have been initiations of transcription occurring from within Tn1725. Nevertheless, we feel that a large fraction of Tn1725 insertions are expected to result in nif$^-$ phenotype in Azospirillum. Since at least in the case of pMS175-2::Tn1725 insertions, all the Cm^r colonies were Tc^s (Tc^r is a marker present only on the vector) but still many were nif$^+$, it is plausible to conclude that Tn1725 transposes at a high frequency in Azospirillum. We never detected spontaneous Cm^r resistance in 29710 rif at frequencies higher than 10^{-9}.

In the third case, i. e., plasmid pMS188 (see results) and Tn5 insertion in a newly identified nif gene of A. brasilense, we obtained a very high percentage (50 %) of nif$^-$ transconjugants, which showed that Tn5 is better suited than Tn1725 for site-directed mutagenesis in Azospirillum. This is further supported by the results of Perroud et al. (see this volume).

We used a novel combination of nif-gene deletion in combination with resistance-gene-cartridge mutagenesis of Abr nifHDK (e.g. pMS205). This method proved the most successful in obtaining a large number of nif^- mutants all of which were completely defective in nitrogen fixation. These nif^- mutants are ideal for field experiments involving Azospirillum-inoculations because, owing to the deletion of nif-genes, no reversion to nif^+ phenotype can occur. Secondly, since the Kan^r-cartridge cannot transpose to other locations in the genome, no mutation can take place owing to the presence of the cartridge. Thus, the 'cartridge-mutants'are isogenic to the parental wild type strain except for the nif^- and Km^r phenotype. The site-directed nif^- mutants produced using Tn1725 or Tn5 (this study and the accompanying paper of Perroud et al.) suffer from the disadvantage that the transposon can cause undesirable deletion or further transpositions in the genome of Azospirillum, which raises complications when a nif^- mutant and a wild type strain are compared with each other.

With the successful use of site-directed transposon-mutagenesis and cartridge-mutagenesis in *A. brasilense*, it should now be possible to study the cloned genes of this organism with respect to gene organization, function and regulation. We are currently trying similar experiments in *A. lipoferum* strains.

Acknowledgements

We are grateful to Drs. C. Elmerich, V. N. Iyer, R. Simon and Prof. R. Schmitt for providing the plasmids pAB1, pGS9, pSUP202 and Rts1::Tn1725, respectively. The excellent technical assistance of Mrs. M. Ohlraun is greatly appreciated. This work was financially supported by the Bundesministerium für Forschung und Technologie.

References

1. Döbereiner, J. (1984). In: Azospirillum II. Genetics, Physiology, Ecology (Ed. Klingmüller, W.),pp. 9-23, Birkhäuser, Basel.
2. Hartmann, A., Singh, M. and Klingmüller, W. (1983). Can. J. Microbiol. 29, 916-923.
3. Quiviger, B., Franche, C., Lutfalla, G., Rice, D., Haselkorn, R. and Elmerich, C. (1982). Biochimie 64, 495-502.
4. Jara, P., Quiviger, B. Laurent, B. and Elmerich, C. (1983). Can. J. Microbiol. 29, 968-972.
5. Singh, M. and Klingmüller, W. (1983). In: Azospirillum II. Genetics, Physiology, Ecology (Ed. Klingmüller, W.), pp. 47-55, Birkhäuser, Basel.
6. Pedrosa, F.O. and Yates, M.G. (1984). FEMS Microbiol. Letters 23, 95-101.
7. Fahsold, R., Singh, M. and Klingmüller, W. 1985 (This volume).
8. Ruvkun, G.B. and Ausubel, F.M. (1980). Proc. Natl. Acad. Sci. USA 77, 191-195.
9. Simon, R., Priefer, U. and Pühler, A. (1983). Bio/technology, 1, 784-791.
10. Selvaraj, G. and Iyer, V.N. (1983). J. Bacteriol. 156, 1292-1300.
11. Chang, A.C.Y. and Cohen, S.N. (1978). J. Bacteriol. 134, 1141-1156.
12. Singh, M. and Klingmüller, W. (1984). In:Proceedings of the III. International Symposium on nitrogen fixation with non-legumes, Helsinki (In press).
13. Albrecht, S.L. and Okon, Y. (1980). Methods in Enzymology 69, 740-749.
14. Singh, M., Kleeberger, A. and Klingmüller, W. (1983). Mol. Gen. Genet. 190, 373-378.
15. Singh, M. and Klingmüller W. (1984). In: Trends in Molecular Genetics, U. Sinha and W. Klingmüller (Eds.), Spectrum Publishing House, Patna, Delhi (In press).
16. Scolnik, P. and Haselkorn, R. (1984). Nature 307, 289-292.
17. Perroud, B., Bhandari, S.K. and Elmerich, C. (1985). This volume.

COSMID CLONING OF NITROGENASE STRUCTURAL GENES OF AZOSPIRILLUM LIPOFERUM

Raimund Fahsold, Mahavir Singh and Walter Klingmüller
Lehrstuhl für Genetik, Universität Bayreuth
Universitätstr. 30, D-8580 Bayreuth, FRG

Summary: Cosmid gene libraries were established from two strains of *Azospirillum lipoferum* (ATCC 29708 and 29731). The clones carrying the *nif* structural genes (*nif*HDK) were isolated. These genes were found to be clustered and organized in the same order as in *Klebsiella pneumoniae* (*nif*HDK). After introduction into the respective *K. pneumoniae* *nif*$^-$ mutants the cloned *A. lipoferum* *nif*HDK (Ali *nif*HDK) restored nitrogenase activity up to approximately 2 % of wildtype. Keywords: *Azospirillum lipoferum*, gene library, cloning, *nif* structural genes, complementation

Introduction

Hybridization of *K. pneumoniae* *nif*HDK (Kp *nif*HDK) DNA to the DNA of other diazotrophic bacteria had shown that the genes coding for the structural proteins of nitrogenase are highly conserved (1). These studies were extended to *Azospirillum* and regions homologous to Kp *nif*HDK were also detected in various *Azospirillum* strains (2,3). The *nif*HDK operon, located on a 6.7 kb EcoR1 fragment, was cloned from strain *A. brasilense* 29145 (4) and mapped (Perroud et al., this volume).

Despite this strong homology, there are interspecies variations regarding the organization of *nif* structural genes, e. g. *nif*HD is separated from *nif*K in *R. japonicum* (5). Also, the restriction enzyme patterns of these genes are quite different (1,5,6,7).

In preliminary experiments these differences were also found between the *nif* structural genes of *A. brasilense* and *A. lipoferum*. This observation and the differences in GC content (8), carbohydrate metabolism (9, 10) and IAA biosynthetic pathways

Azospirillum III: Genetics, Physiology, Ecology
Ed. by W. Klingmüller

(11) prompted us to clone the nifHDK genes of A. lipoferum with a view to comparing the organization and function of nifHDK genes of A. brasilense and A. lipoferum.

Materials and methods

Bacterial strains and plasmids are listed in table 1.

Table 1 . Bacterial strains and plasmids

Strains	relevant characteristics	source/ref.
A. brasilense (ATCC)		
29145, 29711	wildtype	
A. lipoferum (ATCC)		
29708, 29709, 29731	wildtype	
Escherichia coli		
HB101	Sm^r, r^-, m^-, $recA^-$	(J. Collins, Braunschweig, FRG)
BHB2688, BHB 2690		(12)
Klebsiella pneumoniae		
50231	wt	(C. Kennedy, Sussex)
UNF824	nifD2563::Tn5	(13)
UNF1149	nifK2189::Tn7	(13)
UNF1150	nifH2191::Tn7	(13)
CK2901	nifD2290	(14)

Plasmids		
pSUP106	Tc^r,Cm^r	(R. Simon, FRG)
pAB1	Tc^r	(4)
pSA30	Tc^r	(15)
pV34	Amp^r,Tc^r	(T. Hohn, Basel)
pFS191	Cm^r, Al nifHDK	(this work)
pFS13	Tc^r Al nifHDK	(")
pFS393	Tc^r, Cm^r Al nifHDK	(")
pcosAl708-11,-12	Amp^r	(")
pcosAl731-6,-9	Amp^r	(")

Media: All strains were grown in LB (10g Tryptone, 5g NaCl, 5g yeast extract). For measurement of acetylene reduction activity (16) *K. pneumoniae* strains were grown in NFDM.

Preparation of DNA: Lysis of *Azsopirillum* strains and purification of chromosomal DNA was done essentially as described by Marmur (17). The best lysis was achieved using SDS (0.5%) and proteinase K (50 µg/ml). DNA was further purified by cesium-chloride-ethidiumbromide (CsCl-Ethbr) equilibrium density-gradient centrifugation. Plasmid DNA was isolated as described by Rodriguez and Tait (18). Large quantities of plasmid DNA were obtained by the cleared lysate method followed by a final purification in a CsCl-Ethbr equilibrium density-gradient centrigugation (19).

Nick translation and DNA-DNA hybridization: Preparations of pure restriction fragments were labelled with (α-^{32}P)dCTP by nick translation (19,20). Hybridization was done as previously described (19).

In vitro packaging: The procedures of preparing packaging extracts and of *in vitro* packaging followed the outlines recently published (12,21).

Results

Hybridization of chromosomal DNA with nifHDK genes

In order to compare *nif* structural genes of the *Azospirillum* strains, total DNA from 5 strains was digested with EcoR1, separated by agarose gel electrophoresis, transferred to nitrocellulose filters and hybridized using ^{32}P labelled Abr *nif*HDK isolated from pAB1. Each of the strains showed a distinct pattern of hybridizing fragments (Fig. 1). In most of the *Azospirillum* strains, except *A. brasilense* 29145, two or more fragments hybridized with *nif*HDK DNA. This shows that there are differences in the restriction enzyme patterns of *nif*HDK genes of *A. brasilense* and *A. lipoferum* strains. The hybridizing fragments of the *A. lipoferum* strains were: for strains 29708 and

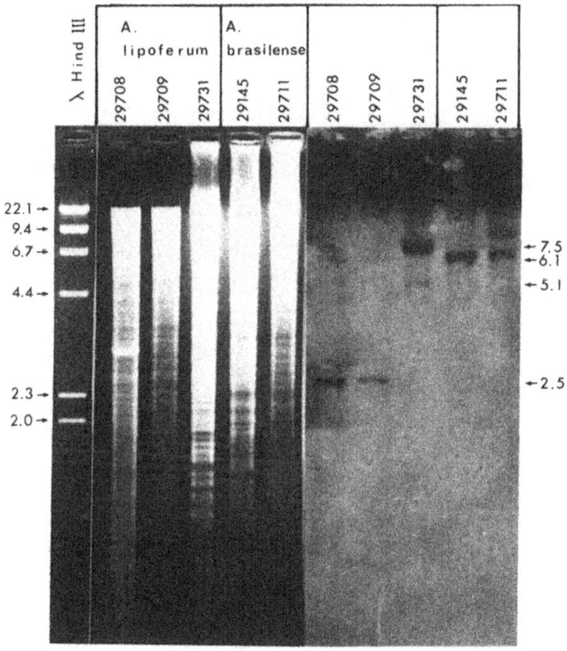

Fig. 1
Detection of nifHDK homologous regions in Azospirillum strains. Southern hybridization between chromosomal DNA of A. lipoferum 29708, 29709, 29731 and A. brasilense 29145, 29711 and ^{32}P labelled Abr nifHDK.

29709, three fragments of 2.5, 5.1 and 6.1kb; for strain 29731, two fragments of 5.0 and 7.5 kb (Fig. 1). Since one fragment in each case hybridized strongly to the probe, it appears that most of the nifHDK is present on one fragment in 29708, 29709 and 29731. The hybridization patterns of strains 29708 and 29709 are very similar.

Construction of gene banks

We decided to clone the nif structural genes of A. lipoferum and concentrated on strains 29708 and 29731. Cosmid gene libraries were established in the cosmid cloning vector pV34. The vector was digested with BamH1 and the total DNA of A. lipoferum was partially digested with Sau3A1. After ligation, the recombinant cosmid molecules were in vitro packaged into phage lambda particles, which were used to infect E. coli HB101. The resulting amp^r colonies were streaked onto nitrocellulose filters for colony hybridization with ^{32}P labelled Abr nifHDK as a probe. From each gene library, 2 clones were found to carry the hybri-

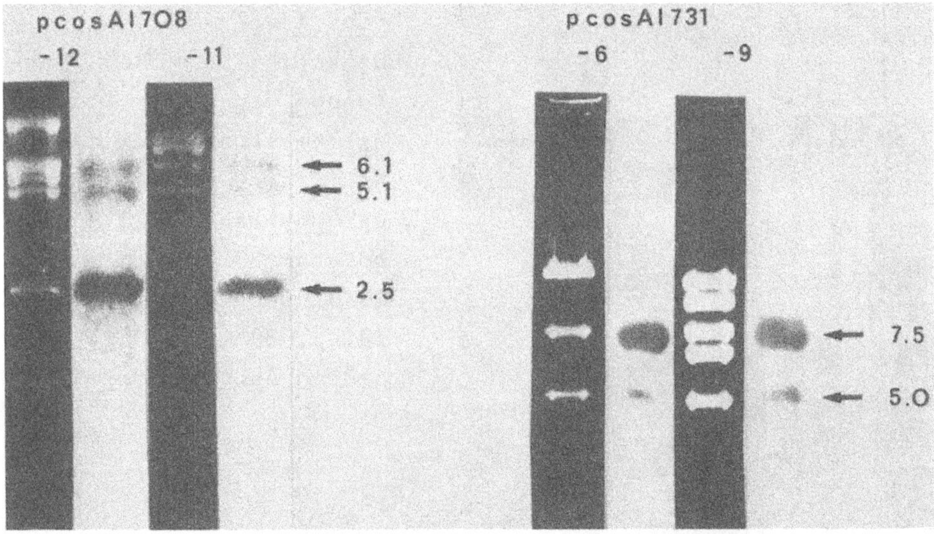

Fig. 2. Localization of nifHDK genes on recombinant cosmids. Southern hybridization between EcoR1 digests of cosmid clones of A. lipoferum and ^{32}P labelled Abr nifHDK. Cosmids pcosA1708-11 and -12 are clones of A. lipoferum 29708 and pcosA1731-6 and -9 those of 29731.

dizing fragments matching well in size with those detected in genomic blots (compare Fig. 1 and Fig. 2). In both cases the cosmids are overlapping and the lengths of the inserts are between 30 and 40 kb.

Subcloning of nif HDK genes

For further mapping and characterization of the nifHDK homologous regions, the respective fragments were subcloned. From pcosA1708-11, a Sal1 partial digest was ligated with Sal1 digested cloning vector pSUP106. After transformation of HB101 the colonies were screened for possession of the whole nifHDK region. From the other A. lipoferum cosmid clone, pcosA1731-6, an EcoR1 partial digest was ligated with the same vector as above, but EcoR1 digested, and recombinant clones were screened for the 5kb and 7.5kb EcoR1 fragments. The plasmids with the subcloned nif structural genes of A. lipoferum 29708 (pFS191) and 29731 (pFS13) were used to determine the order and length of the nifHDK homologous regions. We mapped the inserts by hybridizing single and double digests of pFS191 and pFS13 to parts of the Kp nifHDK operon which were used as the radioactive probes for

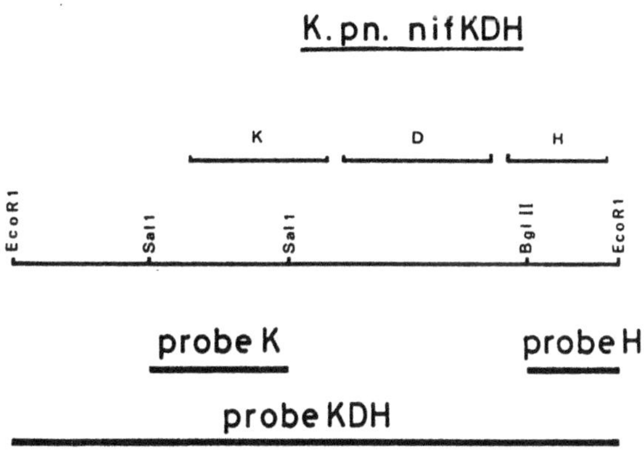

Fig. 3
Map of Kp nifHDK (15)
The probes used for hybridization are indicated by thick bars.

nifH, nifK and nifHDK (Fig. 3). With these probes hybridization was done in parallel and the hybridization patterns generated by each probe were compared. In both cases the length of the nifHDK homologous region was delimited to approximately 5 kb (Fig. 4). In pFS191 this region is located on an 8 kb Pst1 fragment and in pFS13 on two 5.0 and 7.5 kb EcoR1 fragments. The hybridization experiments revealed that the three nif structural genes are present and clustered. The length of the region and order of the genes is similar to nifHDK genes from K. pneumoniae and A. brasilense. A comparison of the restriction enzyme recognition sites on cloned nif DNA of A. lipoferum with that of A. brasilense or K. pneumoniae showed differences. While the nifHDK of the latter organisms are located on a single EcoRI fragment, the respective genes of A. lipoferum have an EcoR1 site within the nifK region. An interesting point is that the genes of the two A. lipoferum strains show strong similarities when one compares the BamH1. EcoRI and HindIII sites in the nifDK region. In the nifH region there are differences if one considers the right

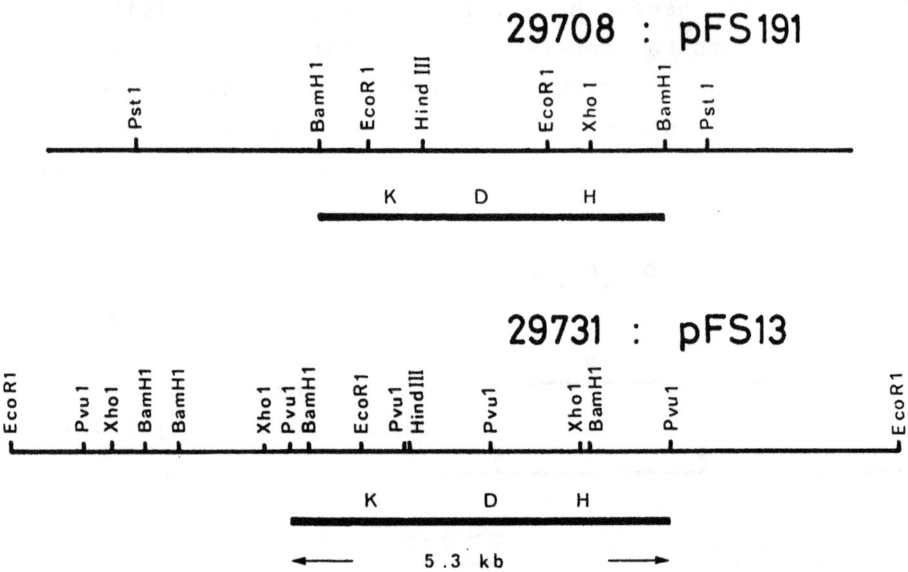

Fig. 4. Physical maps of *nif*HDK genes of *A. lipoferum* strains 29708 (pFS191) and 29731 (pFS13). The *nif* structural genes were delimited by Southern hybridization.

EcoR1 site of the 29708 clone and the right BamH1 site of the 29731 clone. This confirms the close relationship of the *nif* structural genes of the *A. lipoferum* strains studied. Considering the restriction sites mapped in the present study, there are no similarities between the restriction maps of Ali *nif*HDK and the restriction maps of *nif*HDK of *A. brasilense* or *K. pneumoniae*.

Complementation of K. pneumoniae nif⁻ mutants

The cloned *nif* structural genes were tested for complementation of *K. pneumoniae* *nif*H, *nif*D and *nif*K mutants. The *nif*HDK genes of *A. lipoferum* cloned in pFS393 and pFS13 were introduced into *K. pneumoniae* wild type and nif^- mutants. pFS393 carries the 8 kb Pst1 fragment of plasmid pFS191. The transformants were tested for acetylene reduction. All values of acetylene reduction are expressed as % of wildtype activity (Table 2). The *K.*

Table 2. Effect of A. lipoferum nifHDK genes cloned in pSUP106 on acetylene reduction activity of K. pneumoniae wild-type and nif⁻ mutant strains. Values are expressed as % of wildtype activity

Klebsiella strains	Plasmids: no plasmid	pFS393	pFS13
50231 (wt)	100.0	45.4	42.3
UNF1150 (nifH⁻)	0.03	0.1	1.1
UNF824 (nifD⁻)	0.01	2.4	0.8
CK2901 (nifD⁻)	0.01	1.9	2.0
UNF1149 (nifK⁻)	0.06	0.5	0.4

pneumoniae mutants containing the cloned A. lipoferum genes showed a low level of reduction of acetylene reaching values of up to approximately 2 % of wild type activity. These values are comparable to those reported for cloned Azotobacter chroococcum genes where complementation of K. pneumoniae nif⁻ mutants reached values up to 3 % of wild type (22). Interestingly, we could not find any complementation with the cloned genes of A. brasilense. A slight inhibition of acetylene reduction in the K. pneumoniae wild type was observed after introduction of the cloned A. lipoferum nifHDK genes on pSUP106 (Table 2). This effect was also observed in experiments with cloned Kp nifHDK genes on multicopy plasmids (23).

Discussion

The restriction maps established for the nifHDK genes of two A. lipoferum strains showed similarities and differences. The similarities are, for example, the three restriction enzyme sites BamH1, EcoR1 and HindIII in the nifDK region. There are differences in the nifH region, for example, the right EcoR1 site on pFS393. Analogous to the genes of K. pneumoniae, the three *nif* structural genes are clustered and the order is *nif*HDK. Further mapping with additional restriction enzymes is necessary for detailed comparison. The cloned genes of *A. lipoferum* restored nitrogenase activity of the respective *K. pneumoniae* *nif*$^-$ mutants. Since the experiments were done with the *A. lipoferum* genes cloned in one direction only, the question is still open whether this weak expression of the *nif* structural genes of *A. lipoferum* in *K. pneumoniae* is due to transcription originating from their own promotor or due to a readthrough from a promotor outside of the *A. lipoferum* DNA insert, i.e. from the chloramphenicol acetyltransferase or tetracyline resistance promotors on the vector. This can be investigated by inverting the insert in the vector followed by complementation experiments. A possible *nif*A gene product titration by the promotor of *A. lipoferum* *nif*HDK is indicated by the decreased values of acetylene reduction in *K. pneumoniae* wildtype containing plasmids pFS393 or pFS13. Currently, we are investigating whether this effect can be abolished by introduction of Kp *nif*A gene on a multicopy plasmid. A positive effect of extra copies of *nif*A should also increase the values of complementation as was shown for the *Azotobacter chroococcum* cloned *nif*HDK genes (22).

Acknowledgements

We are grateful to Drs. R. Simon and C. Elmerich for providing the plasmids pSUP106 and pAB1, respectively. We also sincerely thank Dr. C. Kennedy for supplying the *K. pneumoniae* strains. We greatly appreciate the financial support of the Bundesministerium für Forschung und Technologie.

References

1. Ruvkun, G. B. and Ausubel, F. M. (1980). Proc. Natl. Acad. Sci. USA 77, 191-195
2. Elmerich, C. and Franche, C. (1982). In Azospirillum: Genetics, Physiology, Ecology. W. Klingmüller (Ed.) Birkhäuser Verlag, Basel, Experientia Supplementum 42, 9-17
3. Wenzel, W., Singh, M. and Klingmüller, W. (1983). In Azospirillum II: Genetics, Physiology, Ecology. W. Klingmüller (Ed.), Birkhäuser Verlag, Basel, Experientia Supplementum 48, 39-46
4. Quiviger, B., Franche, C., Lutfalla, G., Rice, D., Haselkorn, R. and Elmerich, C. (1982). Biochimie 64, 495-503
5. Fischer, H.-M. and Hennecke, H. (1984). Mol. Gen. Genet. 196, 537-540
6. Haselkorn, R., Golden, J., Robinson, S. J., Avtges, P., Kranz, R. G. and Scolnik, P. A. (1985). In Nitrogen fixation and CO_2 Metabolism. P.W. Ludden and J. E. Burris (Eds.), Elsevier Science Publishing Co., Inc., pp 83-90
7. Adams, T. H., McClung, C. R. and Chelm, B. K. (1984). J. Bacteriol. 159, 857-862
8. Tarrand, J. J., Krieg, N. and Döbereiner, J. (1978). Can. J. Microbiol. 24, 967-980
9. Martinez-Drets, G., Del Gallo, M., Burpee, C. and Burris, R. H. (1984). J. Bacteriol. 159, 80-85
10. Goebel, E.M. and Krieg, N. R. (1984). J. Bacteriol. 159, 86-92
11. Hartmann, A., Singh, M. and Klingmüller, W. (1983). Can. J. Microbiol. 29, 916-923
12. Hohn, B., (1979). In Methods in Enzymology 68. R. Wu (Ed.), Academic press, New York, London, Toronto, pp. 299-309.
13. Merrick, M., Filser, M., Kennedy, C. and Dixon, R. (1978). Mol. Gen. Genet. 165, 103-111
14. Dixon, R., Kennedy, C., Kondorosi, A., Krishnapillai, V. and Merrick, M. (1977). Mol. Gen. Genet. 157, 189-198

15. Riedel, G. E., Ausubel, F. M. and Cannon, F. C. (1979). Proc. Natl. Acad. Sci. USA 76, 2866-2870
16. Cannon, F. C., Dixon, R. A., Postgate, J. R. and Primrose, S.B. (1974). Mol. Gen. Genet. 174, 59-66
17. Marmur, J. (1961). J. Mol. Biol. 3, 208-218
18. Rodriguez, R. L. and Tait, R. C. (1983). Recombinant DNA techniques. An introduction. Addison-Wesley publishing company, London
19. Singh, M. and Klingmüller, W. In: Trends in molecular genetics. U. Sinha and W. Klingmüller (Eds.) Spectrum Publishing House, Patna, Delhi, pp. 15-34, (in press)
20. Rigby, P.W.J., Dieckmann, M., Rhodes, C. and Berg, P. (1977). J. Mol. Biol. 113, 237-251
21. Singh, M. and Klingmüller, W. (1984). In Proceedings of the 3rd international symposium on nitrogen fixation with non-legumes, Helsinki, (in press)
22. Jones, R., Woodley, P. and Robson, R. (1984). Mol. Gen. Genet. 197, 318-327
23. Riedel, G. E., Brown, S. E. and Ausubel, F. M. (1983). J. Bacteriol. 153, 45-56

CONSTRUCTION OF A GENOMIC LIBRARY OF AZOSPIRILLUM BRASILENSE SP7 AND CLONING OF THE GLUTAMINE SYNTHETASE GENE

C. Fogher, H. Bozouklian, S.K. Bandhari and C. Elmerich

Unité de Physiologie Cellulaire, Département de Biochimie et Genétique Moléculaire, Institut Pasteur, 28 rue du Dr. Roux 75724 Paris Cedex 15, France

Summary: A gene library of total DNA of *A. brasilense* Sp7 was constructed in the broad host range cosmid vector pVK100. A plasmid, designated pAB44, that restored by complementation a Gln^{+} Nif^{+} phenotype to the Gln^{-} Nif^{-} *A. brasilense* mutant 7029, was isolated. Plasmid pAB44 contained two *Hind*III fragments of 20 and 10 kb. The *glnA* gene was localized on the 20 kb fragment by Tn*5* insertion and DNA hybridization with a *K. pneumoniae glnA* specific probe. Subclones of the 20 kb fragment were obtained in pBR322. No expression of pAB44 was found in *E. coli* or *K. pneumoniae* Gln^{-} mutants. However, plasmid mutants which restored glutamine independent growth were obtained in *E. coli* strain ET8051, that carries a *glnAntrBC* deletion. The glutamine synthetase was also characterized in *E. coli* ET8051 crude extracts. Key words: gene bank construction, *glnA* gene, glutamine synthetase activity, nitrogen fixation control.

Introduction

A major progress in molecular genetics of *Azospirillum* can be achieved with the disponibility of a large collection of mutants and with the availability of a gene library of total DNA in a broad host range vector that allows the identification of cloned genes directly by complementation of *Azospirillum* mutants. Plasmids such as RK2, RP4, RSF1010 or their derivatives with broad host range (1) can be used as vectors in such experiments. We previously cloned *Azospirillum nif* DNA in pRK290 (2) and stable *nif* partial diploids were obtained (3). Knauf and Nester (4) constructed the pVK100 cosmid vector, a derivative of pRK290, which confers resistance to kanamycin (Km) and tetracycline (Tc). It contains a

Azospirillum III: Genetics, Physiology, Ecology
Ed. by W. Klingmüller

single HindIII site in the kanamycin marker. This vector is not self-transmissible, but it can be transferred by complementation in trans with the Tra functions of P type plasmid (5).

We report here the construction of a gene library of A. brasilense Sp7 composed of HindIII restricted DNA fragments cloned in pVK100 and the identification of a clone containing the glutamine synthetase gene.

Materials and methods

Bacterial strains and plasmids

Table 1 Strains and Plasmids

Strain		Genotype or Phenotype	Reference
Azospirillum brasilense			
Sp7		wild type	(6)
7029		Gln^- Nif^-	(7)
Escherichia coli			
S17.1		pro, thi, recA, (RP4-2-, Tc::Mu-Km::Tn7, Tra^+,IncP-1)	(5)
C600		trp, his, recA, rif	
ET8051		deletion glnAntrBC	(8)
CSR603		maxicell strain	(9)
Plasmid	**Vector**	**Phenotype**	**Reference**
pVK100		Mob^+, Km^R, Tc^R, cos	(4)
pRK2013		Tra^+, Tc^R	(2)
pGE10	R68-45	Tra^+, Km^R, Tc^R, IncP-1, glnAntrBC regulon of K. pneumoniae	(10)
pGE100	pACYC184	Cm^R glnAntrBC regulon of K.pneumoniae	(10)
pGE102	pACYC184	Cm^R glnA of K. pneumoniae	(10)
pAB44) pAB221) pAB223)	pVK100	Tc^R, Sp7 DNA fragments	this work
pAB441	pVK100	Tc^R, pAB44 20 kb HindIII fragment	see Figure 2
pAB442	pVK100	Tc^R, pAB44 10 kb HindIII fragment	"
pAB444	pBR322	Amp^R, pAB441 9.5 kb HindIII-EcoRI fgt	"
pAB444B		mutant derivative of pAB444	
pAB445	pBR322	Amp^R, pAB441 5.2 kb HindIII-EcoRI fgt	"
pAB446	pBR325	Amp^R, pAB441 4.3 kb EcoRI fragment	"

Media

Complete and minimal media for _E. coli_ and _A. brasilense_ have been described previously (7, 11, 12).

DNA purification and electrophoretic analysis.

Total DNA was purified according to Quiviger _et al_ (13). Plasmid DNA was prepared according to Humphreys _et al_ (14) and small scale preparation was performed by the method of Birnboim and Doly (15). Restriction analysis and gel electrophoresis were as previously described (13)

Construction of a gene library

Total DNA of _A. brasilense_ Sp7 was partially digested with _Hin_dIII. After electrophoresis, the DNA fragments whose size ranged between 14 and 30 kb were electro-eluted. Vector DNA digested with _Hin_dIII and treated with calf intestinal alkaline phosphatase was mixed to the DNA fragments. Ligation was carried out overnight at 16°C and the mixture was incubated with _in vitro_ packaging extracts prepared by the method of Hohn (16). The phage particles were adsorbed onto _E. coli_ S17.1 and the infected cells were plated on complete medium containing 10 µg/ml Tc.

Enzyme assay

Nitrogenase was assayed on whole cells under derepression conditions as described earlier (7). Glutamine synthetase (GS) transferase activity was measured as described by Bender (17). Non-denaturing protein-polyacrylamide gel electrophoresis was conducted as SDS-PAGE but SDS was absent from all buffers (18). Gels were assayed _in situ_ for GS transferase activity qualitatively.

Results

Analysis of the gene library

The gene library of _Azospirillum_ Sp7 was composed of 1600 Tc^R Km^S clones. Analysis of 100 clones by _Hin_dIII-_Eco_RI digestions showed that less than two percent of the clones were identical and that the average size of the insert was 23 kb. A typical example of the restriction pattern of a few clones is shown in Figure 1.

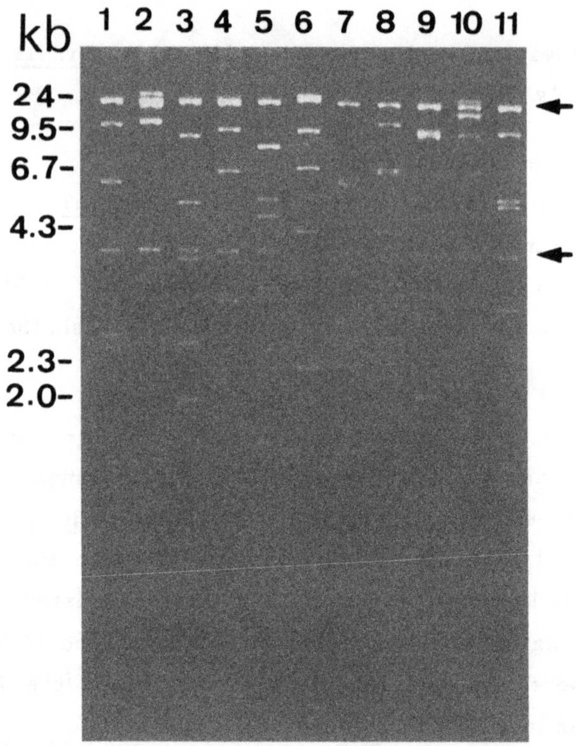

Figure 1. Restriction analysis of 10 randomly choosen clones from the library. 1 to 5 and 7 to 11: plasmids digested with HindIII plus EcoRI; 6: bacteriophage lambda DNA digested with HindIII. The two arrows indicate the pVK100 vector fragments.

Isolation of the glutamine synthetase clone

Using a quick mating procedure (19), we analyzed 250 clones for complementation of the glutamine auxotroph 7029. Three clones, containing plasmid pAB44, -221 and -223, were found to restore glutamine independent growth of strain 7029. Restriction analysis showed that they share in common a 20 kb HindIII fragment. The 3 plasmids and the pVK100 vector were transferred by conjugation to strain 7029 and the conjugation mixture was plated on minimal medium containing 10 μg/ml Tc and supplemented or not with 200 μg/ml glutamine. As shown in Table 2, with pVK100 no transconjugants were observed in the absence of glutamine. For the recombinant plasmids, the frequency of colonies selected was of the same order of magnitude on both media. In addition all the colonies selected on the glutamine containing medium were able to grow in the absence of glutamine. These results strongly suggested that the ability to restore glutamine independent growth in 7029 was a property of pAB44, -221 and -223.

Table 2 Complementation of *A. brasilense* 7029 for Gln^+ phenotype

Donor	Recipient	Transconjugant frequency on selective medium - glutamine	+ glutamine
S17-1(pVK100)	7029	$< 0.8\ 10^{-8}$	$0.6\ 10^{-5}$
S17-1(pAB44)	7029	$0.3\ 10^{-5}$	$1.5\ 10^{-5}$
S17-1(pAB221)	7029	$0.7\ 10^{-6}$	$0.7\ 10^{-6}$
S17-1(pAB223)	7029	$0.5\ 10^{-6}$	$0.4\ 10^{-6}$

Transconjugants were selected on minimal medium for *Azospirillum* with or without 200 µg/ml glutamine and supplemented with 10 µg/ml Tc.

Table 3 Nitrogenase activity of *A. brasilense* 7029 containing various pVK100 derivatives.

	Nitrogenase specific activity nmol C_2H_4 x min^{-1} x mg $protein^{-1}$
Sp7	60
7029	1.6
7029(pAB44)	73
7029(pAB221)	90
7029(pAB223)	62
7029(pGE10)	50
7029(pAB441)	70
7029(pAB442)	1.2

Nitrogenase activity was assayed under derepression conditions. Cells were incubated in nitrogen-free medium under an atmosphere of $Ar-O_2$ (99.5-0.5%)

Strain 7029 has a Gln^- Nif^- phenotype and we measured the nitrogenase activity in the different transconjugants (Table 3). Cosmid pVK100 did not modify the nitrogenase activity of strains 7000 and 7029 (data not shown). Plamids pAB44, -221 and -223 that restored glutamine independent growth restored a normal control of nitrogen fixation in 7029. Growth in the absence of glutamine and nitrogen fixation of 7029 were also restored by pGE10, which contains the *glnAntrBC* regulon of *K. pneumoniae* (Table 3). Introduction, in strain 7029 containing pGE10, pAB44, -221, or -223, of pPH1 (20) which is incompatible with other IncP-1 plasmids, led to the recovery of the Gln^- Nif^- phenotype (data not shown).

Subcloning of the *glnA* region from cosmid pAB44

The 20 and 10 kb *Hin*dIII fragments contained in pAB44 were subcloned in pVK100, and the resulting plasmids pAB441 and pAB442 (Figure 2) were transferred into strain 7029. Only pAB441 complemented the mutant (Table 3). Plasmid pAB441 *Hin*dIII-*Eco*RI fragments were subcloned in pBR322 and in pBR325 to yield pAB444, -445, and -446 (see Figure 2 and Table 1).

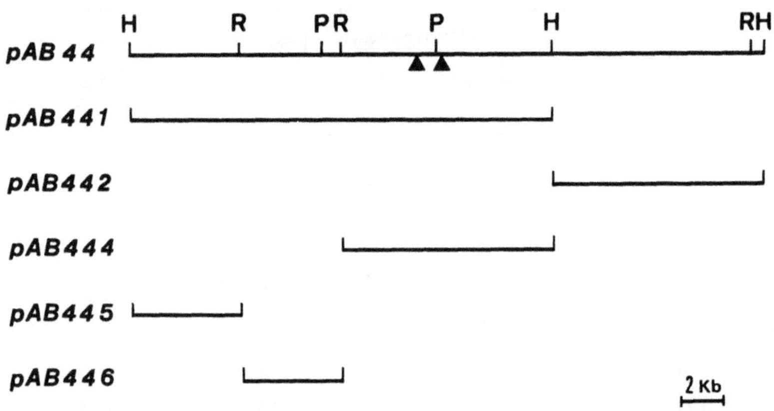

Figure 2. Physical map of the *Azospirillum* DNA cloned in pAB44 and derivatives. Restriction sites: H: *Hin*dIII; R: *Eco*RI; P: *Pst*I. Black triangles indicate the position of Tn*5* insertions which abolished complementation of mutant 7029.

The plasmids pAB441, -444, -445 and -446 were introduced in E. coli ET8051 and no complementation for glutamine independent growth was observed. However, after four days of incubation, glutamine independent mutants appeared on the plates at a frequency of 10^{-7} in transformants containing pAB441 and -444. The mutation was plasmid-borne, since after plasmid extraction from a few mutants, followed by transformation of ET8051, 100 percent of the transformants were able to grow after overnight incubation on minimal medium. Plasmid mutants displayed the same restriction pattern as the original clones. The derivative of pAB441 still restored growth and nitrogen fixation when introduced into 7029. The derivative of pAB444, designated pAB444B, was further used to characterize the GS activity in crude extract of ET8051.

Hybridization with a K. pneumoniae glnA specific probe revealed only the 10 kb HindIII-EcoRI fragment (see Figure 3).

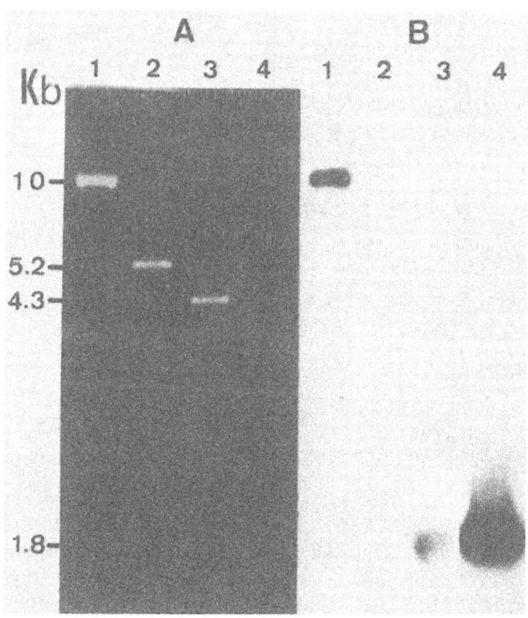

Figure 3. Hybridization of the K. pneumoniae glnA probe to DNA fragments of 10 (lane 1), 5.2 (lane 2) and 4.3 (lane 3) kb purified from pAB441. A: agarose gel electrophoresis; B: corresponding autoradiogram after hybridization with the probe. Lane 4: 1.8 kb SmaI fragment used as probe, and purified from pGE102, which contains a part of the K. pneumoniae glnA gene.

To localize more precisely the glnA gene, we looked for Tn5 insertions in pAB441 which abolished complementation of 7029. About 200 clones containing a Tn5 insert in pAB441 were crossed with 7029. Two clones, which did not restore the Gln^{+} phenotype, were identified. In the corresponding plasmids, Tn5 was localized in the 10 kb HindIII-EcoRI fragment, near the unique PstI site (Figure 2). The two plasmids did not gave rise to glutamine independent mutants when introduced into ET8051.

Characterization of the glutamine synthetase product.

Extracts of ET8051 containing pAB444B and pGE100 were subjected to electrophoresis under non-denaturing conditions in 5 percent polyacrylamide gel (see Figure 4).

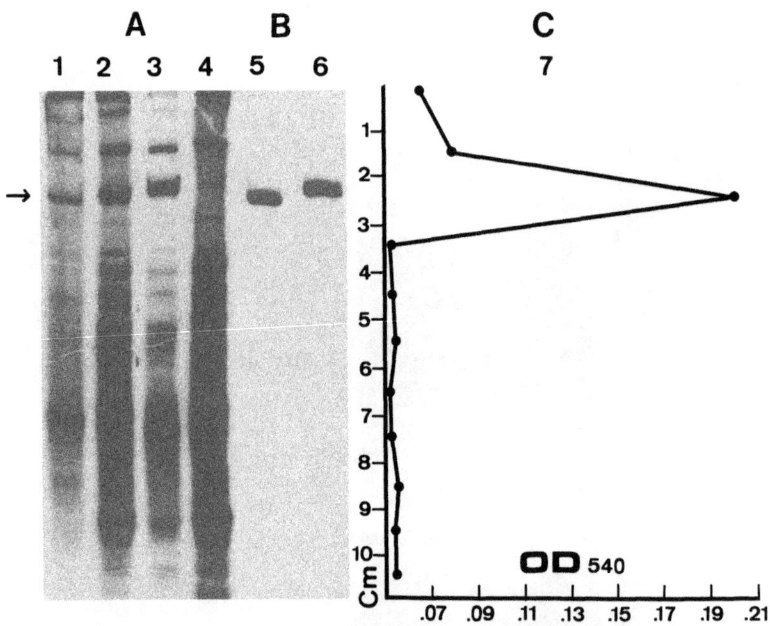

Figure 4. Non-denaturing polyacrylamide gel electrophoresis of protein extracts of strains ET8051 and CSR603 containing pAB444B or pGE100. A: Coomassie blue staining 1: CSR603(pAB444B); 2: ET8051(pAB444B); 3: ET8051(pGE100); 4: ET8051 without plasmid. B: Autoradiogram of labelled protein extract of CSR603 containing: 5: (pAB444B); 6: (pGE100). C: Determination in situ of GS activity in extracts of ET8051(pAB444B). Gel slices of 1 cm were incubated in the GS assay mixture and the OD 540 was measured.

Non-denaturing gel electrophoresis revealed a single transferase activity-staining band corresponding to the protein band indicated by an arrow in Figure 4A (data not shown). Correspondence between the activity-staining band, the radioactive band (Figure 4B), and the activity peak (Figure 4C), clearly identify the GS product. The apparent molecular weight of the GS subunit was found to be 50 kd, after SDS gel electrophoresis of maxicell products encoded by pAB444B (data not shown).

Discussion

Results reported here establish that plasmid pAB44, isolated from the gene library by complementation of strain 7029, contains the glutamine synthetase structural gene, glnA, of Azospirillum Sp7. Evidence was as follows:

i) Hybridization of pAB44 with total DNA of Sp7 digested with HindIII plus EcoRI revealed bands of the same size as the cloned fragments (data not shown). Thus the insert DNA in pAB44 originated from Azospirillum.

ii) pAB44 10 kb HindIII-EcoRI fragment shared homology with a K. pneumoniae glnA probe. The gene was not localized precisely on this fragment, but two plasmid derivatives containing a Tn5 insertion near the PstI site failed to complement strain 7029. It is therefore likely that glnA is located in this region.

iii) Complementation of a glnA deletion strain of E. coli was observed with the mutant derivative pAB444B. The absence of complementation of ET8051 by pAB44 or pAB441 suggests that the Azospirillum transcription or traduction signals are not recognized in E. coli.

iv) The glutamine synthetase product was identified on non-denaturing gels.

In E. coli, Salmonella and Klebsiella the glnA gene belongs to a complex regulon (10, 21). This regulon contains the ntrBC genes, the products of which are responsible for the transcriptional activation of a number of operons involved in nitrogen assimilation. For example, in K. pneumoniae nitrogen fixation is under the control of the ntrBC genes. Therefore, some of the glutamine auxotrophs of K. pneumoniae have a Nif^-

phenotype, as a consequence of a mutation in the *ntrBC* genes or as a result of a polar effect of *glnA* mutation on *ntrBC* expression. In *Azospirillum*, Pedrosa and Yates (22) have recently isolated Nif$^-$ mutants of Sp7 whose nitrogen fixation was restored by plasmid pGE10. The authors proposed that the *Azospirillum* mutants were analogous to *K. pneumoniae* *ntrC* mutants. Mutant 7029 of Sp7 has a Gln$^-$ Nif$^-$ phenotype, and thus resembles *glnA* or *ntrBC* mutants of *K. pneumoniae*. Moreover, plasmid pAB44 and pGE10 restored a Gln$^+$ Nif$^+$ phenotype in this mutant. It is thus tempting to speculate that plasmid pAB44 carries the equivalent of the *ntrBC* genes. However, pAB44 did not complement the mutant strains isolated by Pedrosa and Yates (data not shown). This raises the question of the localisation of the *ntrBC* genes in *Azospirillum*. Further experiments are in progress to answer this question.

Acknowledgements

C.F. was the recipient of a postdoctoral fellowship from research contracts funds, S.K.B. was a postdoctoral fellow from the French Government.

This work was supported by funds from the University Paris VII, and by a Research Contract from Elf Bio-Recherche, Entreprise Minière et Chimique, Rhône-Poulenc Recherche, and CDF Chimie.

References

1. Bagdasarian, R.L., Lurz, R. Rückert, B., Franklin, F.C.H., Bagdasarian, M.M., Frey, J. and Timmis, K.N. (1981). Gene 16, 237-247

2. Ditta, G., Stanfield, S., Corbin, D. and Helinski, D. (1980). Proc. Natl. Acad. Sci. USA 77, 7347-7351

3. Jara, P., Quiviger, B., Laurent, P. and Elmerich, C. (1983). Can. J. Microbiol. 29, 968-972

4. Knauf, V.C. and Nester, E.W. (1982). Plasmid 8, 45-54.

5. Simon, R., Priefer, U. and Pühler, A. (1984). BIO/TECHNOLOGY 1, 784-791.

6. Tarrand, J.J., Krieg, N.R. and Döbereiner, J. (1978). Can. J. Microbiol. 24, 967-980

7. Gauthier,D. and Elmerich, C. (1977). FEMS Microbiol. Lett. 2, 101-104

8. Tuli, R., Fisher, R. and Haselkorn, R. (1982). Gene 19, 109-116

9. Sancar, A. A., Hack, M. and Rupp, W.D. (1979). J. Bacteriol. 137, 692-693

10. Espin, G., Alvarez-morales, A., Cannon, F., Dixon, R. and Merrick. M. (1982). Mol. Gen. Genet. 186, 518-524

11. Elmerich, C., Houmard, J., Sibold, L., Manheimer, I., and Charpin, N. (1978). Mol. Gen. Genet. 165, 181-189

12. Franche, C. and Elmerich, C. (1981). Ann. Microbiol. (Inst. Pasteur) 132A, 3-19

13. Quiviger, B., Franche, C., Lutfalla, G., Rice, D., Haselkorn, R. and Elmerich, C. (1982). Biochimie 64, 495-502

14. Humphreys, G.O., Willshaw, G.A. and Anderson, E.S. (1975). Biochim. Biophys. Acta 383, 457-463

15. Birnboim, H.C. and Doly, J. (1979). Nuc. Acids Res. 7, 1513-1523

16. Hohn, B. (1979). Methods in Enzymology, 68, 299-309

17. Bender,R.A., Janssen, K.A., Resnick, A.D., Blumberg, M., Foor, F. and Magasanik, B. (1977). J. Bacteriol. 129, 680-685

18. Laemmli, U.K. (1970) Nature, 227, 680-685

19. Franche, C., Canélo, E. Gauthier, D. and Elmerich, C. (1981). FEMS Microbiol. Lett. 10, 199-202

20. Ruvkun, G. and Ausubel, F.M. (1981). Nature 289, 85-88

21. Merrick, M.J. (1983). EMBO J. 2, 39-44

22. Pedrosa, F. and Yates, M.G. (1984). FEMS Microbiol. Lett. 23, 95-101

CLONING OF THE GENE FOR THE RESTRICTION ENZYME AbrI FROM AZOSPIRILLUM BRASILENSE ATCC 29711

G. Schwabe and W. Klingmüller
Lehrstuhl für Genetik, Universität Bayreuth
Universitätsstraße 30, D-8580 Bayreuth, FRG

Summary: Screening of different Azospirilla strains for restriction/modification enzymes showed, that *A. brasilense* 29711 harbors the class II restriction enzyme *Abr*I and a corresponding modification enzyme. *Abr*I was characterised as an isoschizomer of *Xho*I (1), which cleaves double stranded DNA at the nucleotide sequence 5'-C↓TCGAG-3'. The gene for the restriction enzyme *Pae*R7, which is an isoschizomer of *Xho*I, has been cloned (2) and it is shown here that it hybridizes to specific DNA-fragments of *A. brasilense* 29711. This DNA was used to screen a cosmid gene library from *A. brasilense* 29711 for the *Abr*I gene. One clone, pGBS53, yields a 8 kb *Eco*RI fragment, which produces a strong hybridization signal. Keywords: Restriction enzyme, *Xho*I isoschizomer, gene library, molecular cloning.

Introduction

Restriction enzymes gained much interest as tools for the isolation and characterisation of genes and specific DNA-fragments. Moreover these enzymes and the corresponding sequence specific methyltransferases (modification enzymes) are of more general interest, as they might protect a bacterium against foreign DNA and by this might ensure the genetic identity of this particular bacterium.

To contribute to the general knowledge of the molecular genetics of *Azospirillum*, we started to screen different Azospirilla strains for restriction/modification enzymes.

Azospirillum III: Genetics, Physiology, Ecology
Ed. by W. Klingmüller

Material and Methods

1. Bacteria and DNA

The following Azospirilla strains were used: *A. brasilense* 29145, 29710, 29711 and *A. lipoferum* 29707, 29708, 29731 were from the American Type Culture Collection. *A. amazonense* QRZ42 was from Döbereiner (3). Plasmids pBR322, pACYC177 and pV34 (J. Collins, Braunschweig, FRG) were isolated from *E. coli* HB101 according to standard procedures. Lambda DNA *cI* 857 *Sam*7 was from Boehringer/Mannheim. Adenovirus 2 (Ad2) DNA was from BRL. Chromosomal DNA from Azospirilla strains was isolated according to standard procedures.

2. Chemicals

Agarose was obtained from Seakem. Restriction enyzmes, calf intestine alkaline phosphatase (CIP), and T4 ligase were from Boehringer/Mannheim and were used according to the supplier's guide lines. Nitrocellulose was obtained from Schleicher & Schüll. Bovine serum albumine (BSA), Ficoll and Polyvenyl-pyrrolidone (PVP) were from Sigma.

3. Cell extract preparation and gelelectrophoresis

From 1 ml overnight culture cell extracts were made according to Gingeras and Brooks (2). 1 - 2 μg DNA was incubated with cell extract equivalent to 1 μg protein in 20 μl volume with 10 mM Tris-HCl pH 7.6, 10 mM $MgCl_2$, and 0, 50, 100, or 150 mM NaCl at 30° C for 1 h. The assay mixture was loaded on an agarose gel in 90 mM Tris, 90 mM borate pH 8.3, 2.5 mM EDTA and run overnight at 30 mA. The DNA was stained in 1 μg/ml ethidium bromide.

4. Isolation of DNA-fragments

DNA cleaved by the appropriate restriction enzyme was separated by agarose gelelectrophoresis. The DNA-band was excised and the DNA was electroeluted, purified by phenol and chloroform and percipitated with ethanol.

5. Cosmid gene library

The cosmid pV34 (source J. Collins, FRG) was digested with BamHI/SalI and BamHI/EcoRI and the cosmid arms were isolated. Chromosomal DNA from *A. brasilense* 29711 was digested with increasing amounts of Sau3A. The digest producing fragments of 30 - 40 kb was treated with calf intestine alkaline phosphatase (4) and ligated to the cosmid arms. The ligation mixture was packaged *in vitro* (5) and used to infect *E. coli* HB101. Colonies containing cosmid clones were selected by ampicillin resistance and sensitivity against tetracylin.

6. Hybridizations

6.1 nick translation

For labelling DNA with ^{32}P-dCTP the nick translation kit of BRL was used. The radioactive dCTP was obtained from Amersham.

6.2 colony hybridization

All hybridizations were done in principal according to Maniatis et al. (4). For colony hybridization sterile nitrocellulose filters were layed on replica plated gene library plates and then incubated for 5 min. each in 10 % SDS; 1.5 M NaCl, O.5 M NaOH; 1.5 M NaCl, 1 M Tris-HCl pH 8.0 and in 2 x SSC (20 x SSC = 175.2 g NaCl, 88.2 g Na-citrat/ltr.). The filters were air dryed and baked at 80° C for 2 h. in a vacuum oven. The hybridization was performed in 5 x Denhardt's solution plus 5 x SSC at 60°. (10 x Denhardt's solution = 2 g BSA, 2 g PVP, 2 g Ficoll, o.37 g EDTA, 1 % SDS, 10 x SCC in 1 ltr).

6.3 dot blot

Cosmid DNA was prepared (6), boilt for 10 min. and put on ice. 5 μl DNA solution was applied to a wet nitrocellulose filter. The filter was dryed and baked. The hybridization conditions were the same as for the Southern transfer hybridization.

6.4 Southern transfer

For Southern transfer agarose gels were exposed to UV irradiation for 15 min. The gels were incubated in 1. 0.25 NHCl for 15 min.; 2. in 1.5 M NaCl, 0.5 M NaOH for 45 min.; 3. in 1.5 M NaCl, 1 M Tris-HCl pH 8.0 for 90 min., and 4. in 2 x SSC for 10 min. The DNA was transferred to nitrocellulose filters as described (4). The hybridization was done in 1 x Denhardt's solution plus, 4 x SSC at 68° C for stringent hybridization or at 60° C for a reduced stringency. After hybridization the filters were washed three times in 2 x SSC, 0.1 % SDS, dryed and exposed to a Kodak X-ray film with an intensifier screen.

Results

1. Screening for restriction enzymes

For screening different *Azospirilla* strains for the presence of restriction enzymes phage lambda DNA was incubated at various salt concentrations with cell extracts made from the strains to be tested (for details see Material and Methods). After agarose gel electrophoresis nonspecific nucleases would result in a diminished lambda DNA band, whereas a restriction enzyme would produce a specific band pattern. *A. brasilense* ATCC 29711 clearly produced two bands (Fig. 1). The corresponding class II restriction enzyme was named *Abr*I. For all other strains no indication for such an enzyme was found. There was one exception, preliminary results hint at a restriction enzyme in *A. amazonense* QRZ42 (7).

The incubation conditions for *Abr*I are 10 mM Tric-HCl, pH 7.6, 50 mM NaCl, 10 mM $MgCl_2$ and 37° C. It is worth noting, that *Abr*I accepts 0-100 mM MgCl without change of the cleavage specifity. At 150 mM NaCl the enzyme is inhibited (data not shown). *Abr*I was partially purified by precipitation with 0-50 % $(NH_4)_2SO_4$ and phosphocellulose chromatography. *Abr*I eluted at 50 mM NaCl.

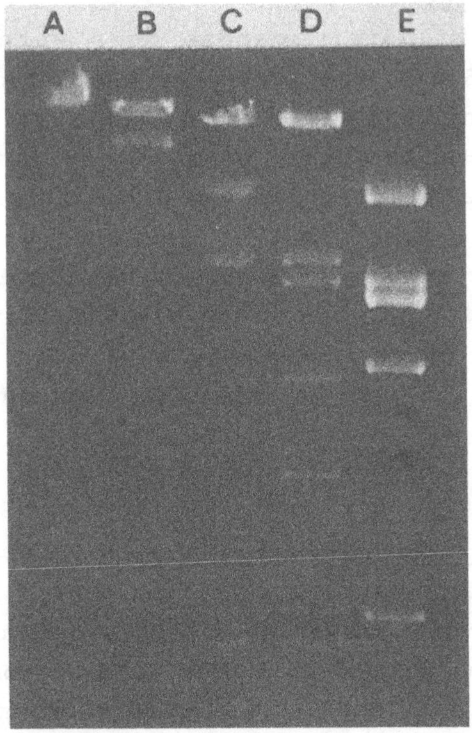

Figure 1. Agarose gel electrophoresis of λDNA and Adenovirus 2 DNA digested with cell extracts of *A. brasilense* 29711 containing *Abr*I.
A= λDNA undigested; B= λDNA digested with *Abr*I; C=λDNA digested with *Hind*III; D= λDNA digested with *Hind*III and *Abr*I; E= Adenovirus 2 DNA digested with *Abr*I. (In the λDNA *Hind*III digest the 4.4 kb fragment appears only as a faint band).

The cleavage site of *Abr*I on the lambda genome was determined by double digests of lambda DNA with *Hind*III/*Abr*I, *Eco*RI/*Abr*I, and *Bam*H/*Abr*I (data not shown). The cleavage site was found to be at approx. 33.5 kb. At 33.493 kb the cleavage site of the restriction enzyme *Xho*I is located (1). On digestion with different DNA species AbrI and XhoI produce identical fragment patterns (data not shown) suggesting that AbrI is an isoschizomer of XhoI.

Next it was examined whether AbrI and XhoI recognize the same nucleotide sequence and in addition cleave the DNA double strand at the same nucleotide within this sequence. The plasmid pACYC177 was cleaved by PstI/AbrI and PstI/XhoI. The two fragments thus generated were purified. A fragment from the PstI/AbrI digest could be ligated with a fragment from the PstI/XhoI digest to produce the complete plasmid. This regenerated plasmid could be recleaved by the parental enzymes. This result indicates that both restriction enzymes produce the same protruding ends. In order to examine whether the first and the sixth base pair of the recognition site is identical for both enzymes, a fragment generated by a digest with SalI (recognition sequence: 5'-G TCGAC-3') was ligated with pACYC177 which was cleaved by AbrI. This ligation produced a plasmid with a hybrid recognition site which could not be recleaved neither by AbrI, XhoI, nor by SalI.

Taken together these results suggest, that AbrI is an isoschizomer of XhoI and cleaves the DNA double strand at

5' - C↓T C G A G - 3'
3' - G A G C T↑C - 5'

2. Modification enzyme in A. brasilense 29711

In most cases a restriction enzyme is accompanied by a corresponding modification enzyme, which protects the chromosomal DNA from digestion. In order to verify that A. brasilense 29711 also harbors a modification enzyme M-AbrI, the chromosomal DNA was isolated and digested by AbrI. As shown in figure 2, AbrI does not cleave the chromosomal DNA of this strain, although the DNA can be cleaved by other restriction enzymes such as BamHI

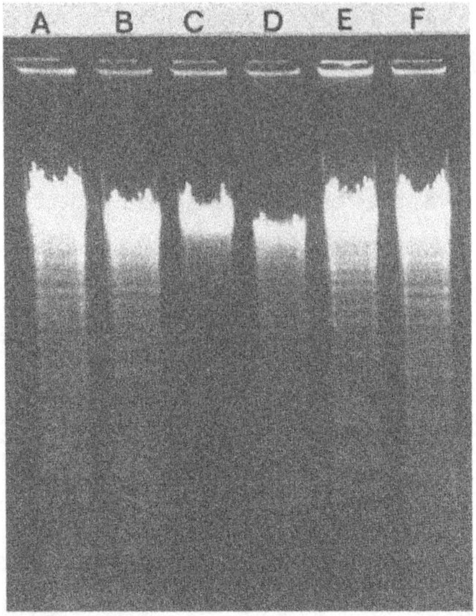

Figure 2. Agarose gel electrophoresis of chromosomal DNA digested with AbrI. A= DNA from *A. brasilense* 29145, B= *A. brasilense* 29710, C= *A. brasilense* 29711; D= *A. amazonense* QRZ42; E= *A. lipoferum* 29707; F= *A. lipoferum* 29708.

(data not shown). AbrI can digest the chromosomal DNA from other Azospirilla strains indicating that the other Azospirilla strains do not have a corresponding modification enzyme (Fig. 2).

The chromosomal DNA from *A. amazonense* QRZ42 is cleaved by AbrI only at a very limited amount. This result is another hint at an restriction/modification system present in *A. amazonense*.

3. Cloning of the AbrI-gene

As a basis for further studies on the molecular genetics of the AbrI restriction/modification system, we started to identify and clone the corresponding genes.

The restriction enzyme PaeR7 from Pseudomonas aeruginosa is an isoschizomer of XhoI too. The gene coding the enzyme has been cloned in the plasmid pPAORM 3.8 (2) and was available as a hybridization probe. Fig. 3 shows, that this DNA hybridizes at stringent conditions to chromosomal DNA isolated from A. brasilense 29711.

The chromosomal DNA isolated from other Azospirilla strains did not produce any hybridization signal neither at high nor at reduced stringency. There was one exception, the DNA from

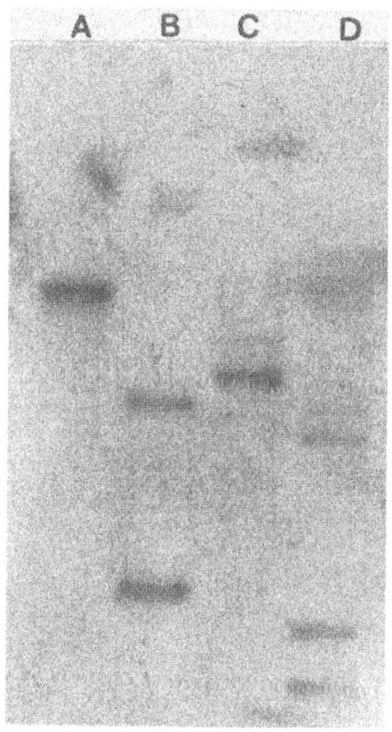

Figure 3. Analysis of chromosomal DNA from A. brasilense 29711 for the restriction enzyme AbrI gene by southern transfer hybridization.
From plasmid pPAORM 3.8 the fragment carrying the gene for Pae7 was radioactively labeled and hybridized with DNA digested with SmaI (lane A), PstI (lane B), EcoRI (lane C), and with BamHI (lane D).

A. amazonense QRZ42 showed a slight hybridization signal at a reduced stringency (7). These experiments prooved, that the PaeR7 gene can be used to screen a gene library for the AbrI genes.

Chromosomal DNA from A. brasilense 29711 was cloned in the cosmid pV34. Approx. 550 colonies, which contained 30 kb inserts, were obtained. This gene library was screened by colony hybridization, producing 95 positive colonies. The cosmid DNAs from these colonies were isolated and used for dot blot hybridization. The DNA from 12 cosmids produced hybridization signals. The corresponding cosmid DNA's were isolated, digested by EcoRI, and analysed by Southern transfer hybridization. Four

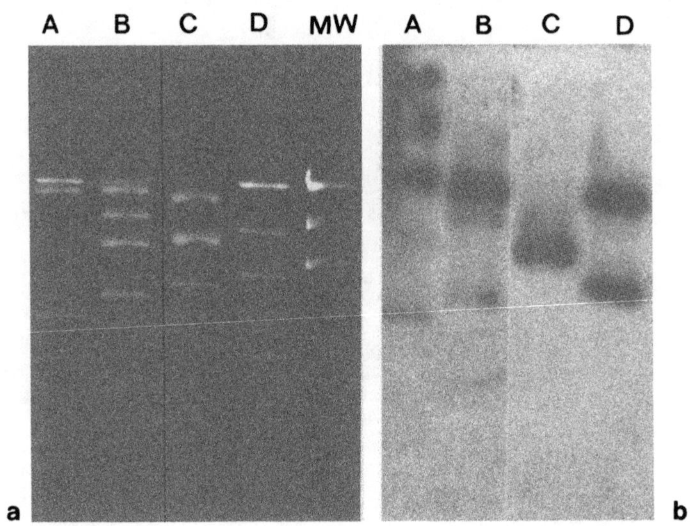

Figure 4. Southern transfer hybridization of cosmid clone pGBS53. Cosmid DNA was digested with SamI (lane A), PstI (lane B), EcoRI (lane C), and with BamHI (lane D). Lambda DNA digested with HindIII was used as a molecular weight standard (MW). ^{32}P-pPAORM3.8 carrying the gene for PaeR7 was used. a) Agarose gel b) Autoradiography of Southern blot.

cosmid clones turned out to be positive and possibly contained the AbrI genes. The Southern transfer analysis of clone pGBS53 is shown in Fig. 4.

As judged from the hybridization experiment and referring to the λHindIII standard the AbrI gene is located on the 8 kb EcoRI-fragment. The intensity of the ethidium bromide staining indicates, that there are two fragments of this size. These fragments are just being subcloned for detailed investigations.

Discussion

In the present report we showed that A. brasilense ATCC 29711 possesses a type II restriction enzyme AbrI and a modification enzyme, which protects DNA from digestion by AbrI. AbrI is an isoschizomer of XhoI, and recognizes the nucleotide sequence 5'CTCGAG3'. Furthermore the corresponding AbrI genes could be cloned. A 8 kb EcoRI fragment from cosmid clone pGBS53 presumably contains these genes. Different strains of A. brasilense, A. lipoferum, and A. amazonense were emploited in these investigations. For A. amazonense QRZ42 there is circumstantial evidence hinting at a restriction enzyme AamI (7). Surprisingly for the other strains no evidence for a restriction enzyme could be obtained under conditions used in this investigation. These findings led us to conclude that the members of the genus Azospirillum do not possess a common restriction/ modification system to ensure their genetic identity. There must be an additional method for the protection against foreign DNA. For some strains a very active nonspecific nuclease could be found. It has a molecular weight of 15 - 20 kilodaltons (unpublished results). The cloned AbrI genes should enable us to answer the question, whether other Azospirilla strains harbor related genes although the corresponding restriction enzyme could not yet be found.

Acknowledgement

We thank Dr. M. Singh for helpful discussions and Dr. T. Gingeras for providing the plasmid pPAORM3.8. Thanks are also due to G. Wirth for excellent technical help and to B. Gubitz for typing the manuscript.

References

1. Gingeras, T.R., Myers, P.A., Olson, J.A., Hanberg, F.A. and Roberts R.J. (1978). J. mol. Biol. 118, 113-122.
2. Gingeras, T.R. and Brooks, J.E. (1983). Proc. Natl. Acad. Sci. USA 80, 402,406.
3. Döbereiner, J. (1983) in: Klingmüller, W. (ed.) Azospirillum: Genetics, Physiology, Ecology. Experientia Supplement 48, 9-23.
4. Maniatis, T., Fritsch, E.F., and Sambrook, J. (1982). Molecular cloning, a laboratory manual. Cold Spring Harbor.
5. Collins, J. (1979) in: ed. Ray Wu, Methods in Enzymology 68, 309-326.
6. Rodriguez, R.L. and Tait, R.C. (1983). Recombinant DNA Techniques. An introduction. Addison-Wesley Publishing Company.
7. Posseckert, G. (1985). Diplomarbeit, Universität Bayreuth.

STABILITY OF BROAD HOST RANGE PLASMIDS IN AZOSPIRILLUM CLONING OF A 5.9 KBP PLASMID FROM A. BRASILENSE RO7

K. Michiels, M. Vanstockem, J. Vanderleyden and A. Van Gool
F.A. Janssenslaboratorium voor Genetica, Kardinaal Mercierlaan 92
B-3030 Heverlee, Belgium

Summary: In order to extend the number of available *Azospirillum* cloning vectors, two strategies were adopted: application of existing vectors with wide host range and development of a new vector based on an *Azospirillum* replicon.

1. Attempts with some broad host range plasmids of the incompatibility groups P1, Q and W to obtain transfer into and stable maintainance in *Azospirillum*, revealed two potential convenient cloning vectors of the inc P1 group: pRK252 and BIN19. We were unable to demonstrate replication of an incQ or incW plasmid in *Azospirillum*.

2. As disposal of a vector compatible with incP1 type vectors is desirable, we cloned a 5,9 kbp plasmid from *A. brasilense* RO7. The plasmid lacks many common recognition sequences for restriction endonucleases and shows no homology with other plasmid or chromosomal DNA from nine *Azospirillum* strains upon hybridization. These properties will facilitate its development into a suitable vector for *Azospirillum*. Keywords: *Azospirillum*, broad host range vector, incompatibility group, plasmid, vector development.

Introduction

Genetic studies in *Azospirillum* can help us to uncover some determinants that play a role in the complicated association of the bacterium with plant roots, to elucidate the regulation of the interaction process and ultimately, to improve the bacteria-plant association in terms of crop yield.

Until now, the genetic tools that proved applicable in *Azospirillum* are rather scarce: Polsinelli *et al*. (1) transferred the incP1 plasmid pRD1 from *E. coli* to *A. brasilense*, Franche *et al*. (2) reported chromosome mobilization by incP1 plasmid R68.45, and Elmerich and Franche (3) obtained Tn*5* mutants by introducing two incompatible plasmids into *A. brasilense* Sp7. Finally,

Azospirillum III: Genetics, Physiology, Ecology
Ed. by W. Klingmüller

attempts to transform A. lipoferum A23 with plasmid DNA were unsuccesfull (4). So far, pRK290 (5), and some related RP4 derivatives are the only potential cloning vectors for Azospirillum. They share the disadvantages of a relatively large size (pRK290 measures about 20 kbp) and a limited array of unique cloning sites (BglII and EcoRI for pRK290). Therefore, the availability of some other more versatile cloning vectors would greatly facilitate cloning work in Azospirillum.

Materials and methods

1. Bacterial strains and plasmids used in this study are summarized in Table I.

Table I. Bacterial strains and plasmids

Strain		Relevant characteristics	Reference and/or source
E. coli	HB101	F^-, hsdS20 (r_B^-, m_B^-), recA13, ara-14 proA2, lacY1, galK2, rpsL20 (Sm^r), xyl-5, mtl-1, supE44, λ^-	(6)
	DH1	F^-, recA1, endA1, gyrA96, thi^-, hsdR17, supE44, relA1, λ^-	(7)/Volckaert G.
	S17-1	[RP4 2-Tc::Mu-Km::Tn7] integrated Tp^r, Sm^r, Pro^-, thi^-, res^-, mod^+	(8)
	JM109	recA1, endA, gyrA96, thi^-, hsdR17, supE44, relA, λ^-, Δ (lac-proAB) [F', traD36, proAB, lacI9 ZΔM15]	(9)

Table I. continued

Strain	Relevant characteristics	Reference and/or source
A. brasilense		
Sp7		(12)/Elmerich C.
7571	nif^- mutant of Sp7	(10)/Elmerich C.
7317	nifD::Tn5 mutant of Sp7	Elmerich C.
Sp107		(11)/Vlassak K.
R07		Dommergues Y./ Vlassak K.
Sp245		Döbereiner J.
$Sp245NR^-$		Döbereiner J.
Sp246		Döbereiner J.
S631		Vlassak K.
A. lipoferum		
Sp59b		Klingmüller W.
SpBr17		(12)/Klingmüller W.

Plasmid	Relevant characteristics	Reference and/or source
RSF1010	Su^r Sm^r Tra^- Mob^+ incQ	(13)
pRK290	Tc^r Tra^- Mob^+ incP1	(5)
pRK2013	Km^r Tra^+ (RK2) ColE1 replicon	(14)
pRK252	Tc^r Tra^- Mob^+ incP1	Ditta (unpublished)/Barnes W.
BIN19	Km^r Tra^- Mob^+ incP1	(15)
pSa4	Cm^r, Sp^r, Km^r, Tra^-, Mob^+ (low) incW	(16)
pSa727	Cm^r, Sp^r, Km^r, Tra^+ Mob^+ (low) incW	(16)
pSa322	Ap^r, Tra^+, Mob^+	(16)
pUC18	Ap^r ColE1 replicon	(9)
pFAP2002		This study
pFAP1004		This study

2. Media and culture conditions

Azospirillum strains were grown on Nutrient agar or MMAB (see accompanying paper in this volume), E. coli strains on L agar, and both were restreaked every month. Antibiotics were added in the following concentrations (μg/ml): Km (25), Tc (10), Cm (25), Sm (15 for E. coli, 150 for Azospirillum), Ap (50).

XG plates for β-galactosidase assay were prepared according to Messing et al. (17). Liquid cultures were grown in LB at 37°C for E. coli, and in YEP at 30°C for Azospirillum strains.

3. Conjugations and selection of transconjugants were performed as described in the accompanying paper by Vanstockem et al. After conjugation, colonies that grew on the selective medium were restreaked two times on the same medium and were shown to contain the transferred plasmid by gel electrophoresis of lysates according to Kado and Liu (18).

4. Plasmid DNA for restriction analysis was prepared by a small scale procedure based on alkaline lysis (19). More purified large-scale plasmid DNA preparations were obtained by applying a crude extract of an alkaline lysate of a 300 ml culture to a CsCl equilibrium gradient or to a preparative 0,7 % agarose gel electrophoresis if a single plasmid band had to be separated from others. After subjecting the gel piece containing the desired band to 3 cycles of freeze-thawing in liquid nitrogen and centrifugation at 13,000 rpm for 1/2 hr in an eppendorf centrifuge, the supernatant was recovered and concentrated by ethanol precipitation.

5. Restriction endonuclease digests were performed under the conditions recommended by the supplier. Dephosphorylation of vector DNA was accomplished by treatment with 1 U E. coli C47 alkaline phosphatase per μg DNA for 30 min at 37°C in an alkaline Tris buffer (pH 8.5). Ligation reactions were overnight at 15°C. Transformation of the ligation reactions to competent cells was according to the $CaCl_2$ method (19).

For Southern blots and hybridizations, the protocol of Silvahy et al. (20) was adopted, except that the last posthybridization wash was in 0,5 x SSC; 0,1 % SDS.

6. All enzymes were purchased from Amersham (U.K.) except for KpnI (BRL).

Results and discussion

1. Broad host range plasmids.

The last few years, several broad host range plasmids have been developed into convenient cloning vectors (5, 16, 21). As these plasmids are able to replicate in several Gram-negative hosts, other than the species from which they were isolated, we tried to transfer and establish some of them into genus *Azospirillum*.

1.1. IncPl plasmids

Although replication of incPl plasmids pRD1 (1) and R68.45 (2) in *A. brasilense* was demonstrated, this does not necessarely implie that all plasmids of the same incompatibility group will behave alike, because small derivatives may have lost functions involved in displaying the broad host range. Three RP4 derived plasmids were first transferred from *E. coli* HB101 to the mobilizing *E. coli* strain S17-1 and, in a subsequent mating, to *A. brasilense* Sp7.

a) pRK290, a plasmid of about 20 kbp, containing unique EcoRI and BglII cloning sites in its Tc^r gene.

b) pRK252, a smaller (± 10 kbp) and hence more convenient cloning vector with a Tc^r gene harboring two SmaI and one SalI cloning sites.

c) BIN19, a pRK252 derivative coding for Km resistance from which an additional 4 kbp of RP4 DNA has been deleted. Moreover, a 440 bp HaeII fragment from phage M13 mp9 (22) containing the alpha-complementing region of the β-galactosidase gene and an array of unique restriction enzyme sites, has been inserted to construct BIN19, so that plasmids harboring an insert can be easily identified on XG plates. BIN19 measures about 10 kbp and also confers Km resistance, which allows convenient selection in an *Azospirillum* host.

All three plasmids established well in Sp7 and were not altered in size, as shown by the plasmid profiles given in Fig. 1.

1.2. IncQ plasmid RSF1010

The broad host range plasmid RSF1010 is able to replicate in a wide range of Gram-negative hosts including *E. coli*, *P. aeruginosa*, *P. putida*, *Azotobacter vinelandii*, *Methylophilus methylotrophus*, *Rhizobium meliloti*, *Agrobacterium tumefaciens*, *Acetobacter xylinum*, *Alcaligenes eutrophus* and *Rhodopseudomonas spheroides* (21). In spite of this, we were unable to obtain *Azospiril-*

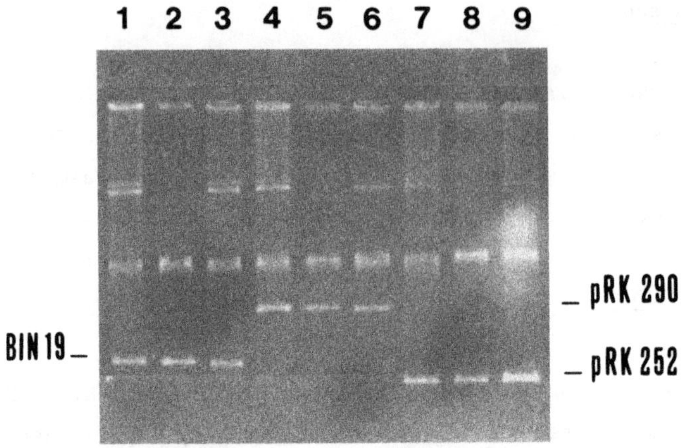

Fig. 1. A. brasilense Sp7 transconjugants carrying BIN19 and pRK290 as shown by Kado and Liu analysis. (1) and (3), Sp7(BIN19); (2), S17-1(BIN19); (4) and (6), Sp7(pRK290); (5), S17-1(pRK290); (7) and (9), Sp7(pRK252); (8), S17-1 (pRK252).

lum transconjugants exhibiting a stable Sm^r phenotype and harboring a RSF1010 plasmid, after a triparental mating HB101 (pRK2013) x HB101 (RSF1010) x Sp7 or SpBr17 wherein pRK2013 served as a helper plasmid to mobilize RSF1010 (21).

1.3. IncW plasmids pSa4 (9,4 kbp) and pSa727 (14,7 kbp)

pSa (29,6 kbp) and its derivative pSa727 have been mobilized into and are functional in a variety of Enterobacteriaceae, Rhizobiaceae and Pseudomonadaceae species (16). Transfer of pSa from Agrobacterium to Azospirillum was reported (23), but pSa was not proved to be present as a replicon. Our attempts with a triparental cross HB101 (pSa322) x HB101 (pSa4 or pSa727) x A. brasilense Sp7 yielded several stable Km^r transconjugants which, however, did not show any additional plasmid. The possibility has to be considered that the plasmids used, which are deletion derivatives of wild-type pSa, exhibit an altered host range. It is likely that the kanamycin resistance trait of the Azospirillum transconjugants is the result of integration of the pSa replicons into the chromosome or indigenous plasmids. This prediction awaits experimental proof.

2. Cloning of a small plasmid from A. brasilense RO7

The results above indicate that obviously only incP1 type broad host range plasmids can replicate in Azospirillum. Therefore we are currently developing a vector based on an Azospirillum replicon.

Several Azospirillum strains were screened for the presence of a small plasmid by the method of Kado and Liu. In A. brasilense RO7 this method revealed a plasmid, approximately 6 kbp in size, together with two larger replicons measuring about 25 kbp and 100 kbp (Fig. 4). The purified DNA of the small plasmid was subjected to an analysis with ten restriction endonucleases, of which surprisingly only KpnI cleaved the DNA into a 5.9 kbp linear molecule (Fig. 2).

As the plasmid to be cloned could be linearized by KpnI, the choice of vectors with a unique KpnI cloning site was obvious: we used pUC18 (2.7 kbp) and pSA4 (9.4 kbp). After ligation into phosphatase treated vector and transformation to E. coli JM109, pUC18 recombinants were selected as white colonies on XG plates containing Ap, whereas pSa4 recombinants were transformed to E. coli DH1 and selected as Cm^rKm^s colonies. A preliminary test

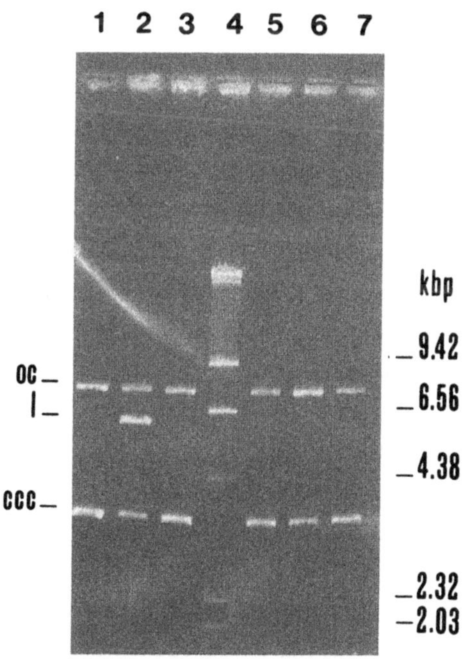

Fig. 2. Restriction analysis of the 5,9 kbp plasmid from strain RO7. (1), EcoRV; (2), KpnI: partial digest yielding 50% linear DNA; (3), PstI; (4), HindIII digest of λ-DNA providing size markers; (5), SalI; (6), XbaI; (7), XhoI.
The plasmid was not cleaved by HindIII, EcoRI or BamHI;(not shown). oc: open circular DNA; ccc: covalent closed circles; l: linear DNA.

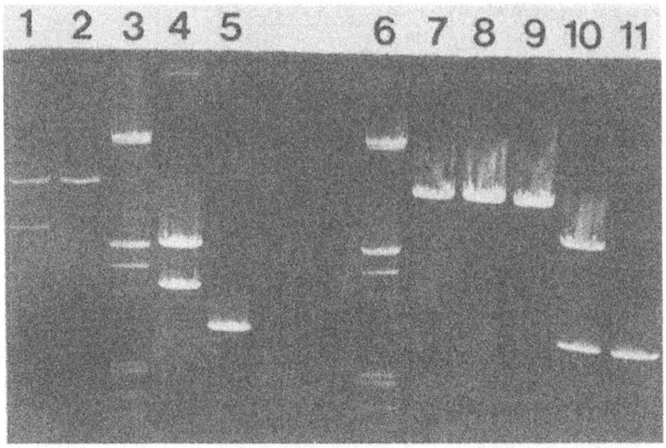

<u>Fig. 3.</u> Restriction analysis of pFAP1004 and pFAP2002.(1), pFAP2002, KpnI; (2), pSa4, KpnI; (3), HindIII/EcoRI combined digest of λ-DNA providing size markers. From bottom to top in kbp: 1,33; 1,58; 1,90; 2,03; 3,53; 4,28; 4,97; 5,15; (4), pFAP 1004, HincII; (5), pUC18, HincII; (6), same as lane 3; (7), pFAP1004, HindIII; (8), pFAP1004, BamHI; (9); pFAP1004, EcoRI; (10), pFAP1004, KpnI; (11), pUC18, KpnI.

had shown that R07 did not produce blue colonies on XG plates, therefore pUC18 recombinants containing an R07 plasmid should not be able to complement β-galactosidase activity in JM109. After purification of several transformants, Kado and Liu analysis revealed the presence of a new plasmid of the appropriate size in both cases (data not shown). Plasmid DNA of one pUC18 recombinant called pFAP1004, and one pSa4 recombinant called pFAP2002 was subjected to restriction analysis for identification (Fig. 3). Both constructions contained the expected 5.9 kbp insert in their KpnI site. The presence of the HindIII, BamHI and EcoRI sites in the polylinker of pFAP1004 was confirmed. Moreover, a HincII recognition site was found to be located in the insert at about 0.9 kbp from the KpnI site. This site will allow us to distinguish clones with alternate orientations of the insert.

As R07, like all <u>Azospirillum</u> strains, is highly resistant to Ap and related antibiotics, we tested <u>E. coli</u> colonies harboring pFAP1004 and pFAP2002 for this trait, but no resistance was expressed.

In pSa4, a single promotor sequence initiates transcription of both Km^r and Sm^r genes, in this order. pFAP2002 is pSa4 carrying an insertion in the

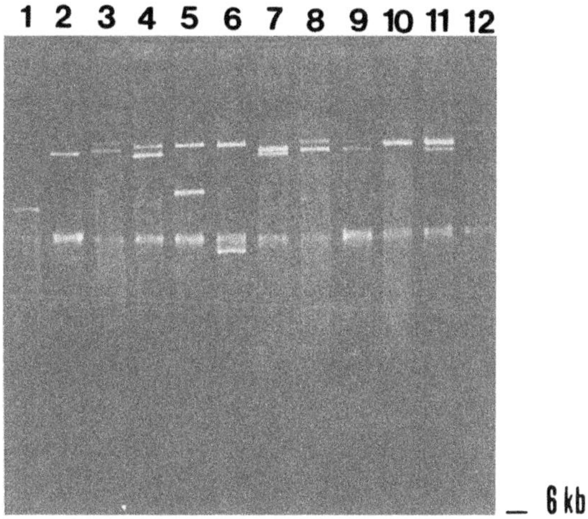

Fig. 4. Plasmid pattern of some Azospirillum strains as prepared by Kado and Liu's method. (1), SpBr17; (2), SpBr14; (3), S631; (4), Sp7; (5), Sp107; (6), R07, showing a faint 5,9 kbp plasmid band. (7), Sp246; (8), 7571; (9), 7317; (10), Sp245; (11), $Sp245NR^-$; (12), Sp59b.
Very large plasmids (+300 kbp) may not be detected by this method because rather stringent lysis is applied.

Km^r gene, thus Sm resistance will be inactivated together with Km resistance, unless a promotor within the insert would drive transcription into the Sm^r gene. pFAP2002 did not confer Sm resistance, indicating the absence of such a promotor. Although this conclusion does not implicitely hold for a clone with inverse insert orientation, it is significant that among 30 independent clones with undetermined insert orientation, none was found to be Sm resistant.

In order to determine eventual homology between the isolated plasmid and other plasmids in strain R07 and in other Azospirillum strains, plasmid profiles (Fig. 4) were transferred to a nitrocellullose membrane and hybridized against a ^{32}P-labelled pFAP2002 probe. In the control experiment, no homology was detected between pSa4 and any of the Azospirillum strains. Under the applied conditions, pFAP2002 hybridized only strongly to the small 5.9 kbp plasmid from R07, not to any other plasmid or chromosomal DNA (data not shown). This lack of homology is a very useful property because it reduces greatly the probability of homologous recombination events between our future vector and host Azospirillum DNA.

In future experiments we will mobilize pFAP2002 into several Azospirillum strains to see whether the R07 replicon is not impaired in replication functions by insertion of pSa4 in its KpnI site. If so, we still dispose of a unique HincII site to reclone the R07 plasmid. It is also to be established whether the new construct is compatible with cloning vectors of the incP1 group. pFAP1004 can be further developed into a convenient shuttle cloning vector for use in E. coli and Azospirillum. A derivative of pFAP2002 will have the advantage of being usable in several other hosts beside E. coli and Azospirillum.

Acknowledgements

We are grateful to Dr. Döbereiner (Brasil), Dr. Elmerich (France), Dr. Klingmüller (Germany) and Dr. Vlassak (Belgium) for providing us with Azospirillum strains.
This research was supported by grants from the "Fonds voor Kollectief Fundamenteel Onderzoek" (F.K.F.O. 2.0013.85) and from the PREST Program (KUL 44). K.M. is aspirant of the N.F.W.O. and M.V. is supported by a fellowship of the PREST Program.

References

1. Polsinelli, M., Baldanzi, E., Bazzicalupo, M. and Gallori, E. (1980). Molec. gen. Genet. 178, 709-711.

2. Franche, C., Canelo, E., Gauthier, D. and Elmerich, C. (1981). FEMS Microbiology Letters 10, 199-202.

3. Elmerich, C. and Franche, C. (1982). In: "Azospirillum, Genetics Physiology Ecology". Klingmüller, W. (ed.), EXS42, pp 9-17, Birkhäuser, Basel.

4. Schwabe, G. and Weber, U. (1983). In: "Azospirillum, Genetics Physiology Ecology". Klingmüller, W. (ed.), EXS48, pp. 56-65, Birkhäuser, Basel.

5. Ditta, G., Stanfield, S., Corbin, D. and Helinski, D.R. (1980). Proc. Natl. Acad. Sci. (USA) 76, 1648-1652.

6. Boyer, H.B. and Roulland-Dunoix, D. (1969). J. Mol. Biol. 41, 459-472.

7. Hanahan, D. (1983). J. Mol. Biol. 166, 557-580.

8. Simon, R., Priefer, U. and Pühler, A. (1983). Biotechnology, nov., 184-190.

9. Yannish-Perron, C., Vieira, J. and Messing, J. (1985). Gene 33, 103-119.

10. Jara, P., Quiviger, B., Laurent, P. and Elmerich, C. (1983). Can. J. Microbiol. 29, 968-972.

11. De-Polli, H., Bohlool, B.B. and Döbereiner, J. (1980). Arch. Microbiol. 126, 217-222.

12. Tarrrand, J.J., Krieg, N.R. and Döbereiner, J. (1978). Can. J. Microbiol. 24, 967-980.

13. Guerry, P., Van Embden, J. and Falkow, S. (1974). J. Bacteriol. 117, 619-630.

14. Figurski, D.H. and Helinski, D.R. (1979). Proc. Natl. Acad. Sci. (USA) 76, 1648-1652.

15. Bevan, M. (1984). Nucleic Acids Res. 12, 8711-8721.

16. Tait, R.C., Close, T.J., Lundquist, R.C., Hagiya, M., Rodriguez, R.L. and Kado, C.I. (1983). Biotechnology, May, 269-275.

17. Messing, J., Gronenborn, B., Müller-Hill, B. and Hofschneider, P.H. (1977). Proc. Natl. Acad. Sci. (USA) 75, 3642-3646.

18. Kado, C.I. and Liu, S.-T. (1981). J. Bacteriol. 145, 1365-1373.

19. Maniatis, T., Fritsch, E.F. and Sambrook, J. (1982). In: "Molecular Cloning", a Laboratory Manual. CSH.

20. Silvahy, T.J., Berman, M.L. and Enquist, L.W. (1984). In: "Experiments with Gene Fusions". CSH.

21. Bagdasarian, M., Lurz, R., Rückert, B., Franklin, F.C.H., Bagdasarian, M.M., Frey, J. and Timmis, K.N. (1981). Gene 16, 237-247.

22. Norrander, J., Kempe, T. and Messing, J. (1983). Gene 26, 101-106.

23. Dubois, M. (1984). Ph.Dr. Thesis. Institut National Agronomique. Paris-Grignon.

TRANSFER AND RANDOM INTEGRATION OF Tn5 IN AZOSPIRILLUM

M. Vanstockem, K. Michiels, J. Vanderleyden and A. Van Gool
F.A. Janssens Laboratory of Genetics, Catholic University
of Leuven, Kardinaal Mercierlaan 92, B-3030 Heverlee

Summary: A method is described for the isolation of at random Tn5 insertion mutants of *Azospirillum spp*. For this purpose a set of plasmid vectors of Tn5, that are suicidal in *Rhizobium* and *Agrobacterium* species, were introduced in *A. brasilense* Sp7 and *A. lipoferum* SpBr17 by conjugation. The plasmids pSUP2021 and pGS9 are suicidal in these *Azospirillum* strains as shown by hybridization experiments of digested total DNA of twenty bona fide Km^r exconjugants with ^{32}P labeled ColE1 :: Tn5. Kanamycin-resistant colonies were detected at a frequency of 10^{-8} per recipient cell. Keywords: *Azospirillum*, transposon mutagenesis, suicide plasmids, Southern hybridization.

Introduction

The initial reports by Döbereiner (1) on the widespread distribution of *Azospirillum* rhizocoenoses has led to considerable interest in determining their potential in agricultural systems. A molecular genetic approach in other plant-interactive microbial systems has been demonstrated to be very successful for the identification of genetic determinants in bacteria-plant interactions. So far only a limited number of tools developed for genetic analysis of Gram negative bacteria have been found to be suitable for *Azospirillum* (for review see 2).

Efforts have been made to perform random mutagenesis of *Azospirillum* using transposon insertions. The advantage of this method consists in the ease by which mutants, induced by insertion of a transposon, can be selected for, due to the antibiotic resistance encoded by the transposon. A major concern in this methodology is to find a suitable plasmid, carrying a transposon with a selectable marker, that can be transferred to but that cannot replicate in *Azospirillum*. Singh (3) and Elmerich (4) have tried for this purpose the plasmid pJB4J1, a plasmid of the incompatibility group P with a

Azospirillum III: Genetics, Physiology, Ecology
Ed. by W. Klingmüller

Mu insert and Tn5 (Km^r). This plasmid has been successfully used to mutagenize Rhizobium. However this plasmid seemed to establish stably in Azospirillum, rendering it not applicable to Azospirillum.

Our attempt was to mutagenize Azospirillum by using two other suicide plasmids, designed for use in Rhizobium and other Gram negative bacteria:

1. The system of Simon et al. (5) consists of a plasmid pSUP2021, carrying Tn5 and a mob site, allowing it to be mobilized with high frequency by the broad host range transfer functions of RP4, integrated in the donor E. coli genome. Since this plasmid is eliminated, unable to replicate in non E. coli hosts, transposition events can be isolated by selecting for Tn5 mediated kanamycin resistance.
2. The system of Selvaraj and Iyer (6) consists of a plasmid pGS9, carrying Tn5, and composed of a p15A-type of replicon and an N-type bacterial mating system, allowing it to be self-transmissible but unable to replicate outside the E. coli host.

The main results of these studies will be given here.

Materials and methods

Bacterial strains and plasmids

The bacterial strains and plasmids used in this study are listed in Table I.

Media and chemicals

For the growth of the plasmid carrying E. coli strains L-broth containing Km was used at 37°C under continuous shaking. Azospirillum spp. strains were grown, at 30°C in yeast extract-peptone broth (YEP). The selection medium for transposon mutants consisted of solidified modified minimal AB medium (MMAB) supplemented with Km. The composition of the MMAB was as follows: K_2HPO_4 (3 g/l), NaH_2PO_4 (1 g/l), NH_4Cl (1 g/l), $MgSO_4.7H_2O$ (0,3 g/l), KCl (0,15 g/l), $CaCl_2.2H_2O$ (0,01 g/l), $FeSO_4.7H_2O$ (0,0025 g/l), Na-malate (5 g/l malate neutralized with NaOH 2,5 N), biotin (0,005 g/l) and microelements [2 ml of a 500 x stock solution, Franche et al. (1981)] (8). The number of acceptor cells was counted on MMAB-agar plates without antibiotics, whereas the donor cells were counted on solid L-agar with Km.

Table I. Bacterial strains and plasmids used

Strain	Relevant properties	Reference and/or source
E. coli		
WA803	Met^- Thi^-	Wood (7)
S17-1	Thi^-, pro^-, $hsdR^+$, $hsdM^+$, recA, Tp^rSm^r [RP4 2-Tc::Mu-Km::Tn7] integrated	Simon et al. (4)
Azospirillum		
A. brasilense Sp7	Ap^r	ATCC 29145
A. lipoferum SpBr17	Ap^r	ATCC 29709
A. brasilense 7317	Km^r nifD::Tn5 Ap^r	Elmerich
Plasmid	**Relevant properties**	**Reference and/or source**
pSUP2021	Col El replicon Ap^r Cm^r Tc^r Km^r(Tn5) tra^- mob^+	Simon et al. (4)
pGS9	p15A replicon Km^r (Tn5) Cm^r tra^+ (N system)	Selvaraj et al. (5)

The concentration of Km used in the solid media, as well as in the liquid LB medium was 25 µg/ml. The antibiotic was obtained from Bristol Laboratories Syracuse N.Y. Chloramphenicol was used at a concentration of 50 µg/ml in solid MMAB to exclude the presence of plasmid cointegrates among potential Tn5 insertion mutants. Chloramphenicol was obtained from Sigma.

Bacterial filter matings

Azospirillum acceptor strains were grown overnight at 30°C in YEP liquid medium (stationary phase). E. coli donor strains were grown overnight at 37°C in $L+Km^{25}$, diluted tenfold in $L+Km^{25}$ and grown another 4 h (log phase). Aliquots (1,5 ml) of donor and acceptor cultures were centrifuged in an eppendorf centrifuge, resuspended and washed once in physiological water. The acceptor cells were submitted to a heat shock (51°C) for 5 minutes prior to mixing with the recipient cells on membrane filters. The mating mixtures were incubated on D-medium (9) for 6 h at 37°C. Filters containing only acceptor or donor cells were incubated for proper controls. The cells were resuspended, washed twice, diluted and spreaded on appropriate plates. After incubation for 2 to 3 days, transposon mutants were purified by streaking for single colonies on MMAB + Km^{25}.

DNA manipulations

Single colonies were picked up and grown overnight in YEP liquid. For the detection of plasmids the bacteria were submitted to lysis according to Kado and Liu (10), the lysates being separated by electrophoresis in 0,7 % agarose gels in Tris-acetate-EDTA buffer (40:2 mM; pH 7,9) at 80V for 6 to 8 h.
Total DNA was prepared as described in the EMBO course on "The use of Ti plasmids as cloning vectors for genetic engineering in plants" (11). DNA concentration of the lysates was determined using λ DNA as a concentration standard (50-100-200 ng/μl) in electrophoresis [0,7 % agarose gels in TB buffer (1:1) at 80V for 1 h]. Restriction digest analysis was performed with Eco RI and Xho I, which were purchased from Amersham and used as recommended by the manufacturer. Gel electrophoresis of the restricted DNA fragments was carried out in 0,8 % agarose gels in TAE buffer for 8 h.

Hybridization technique

^{32}P-labeled DNA of Col E1::Tn5 was prepared by nick translation (Amersham nick translation kit) and was used as a probe in DNA-DNA hybridization. The nitrocellulose sheets (Gene Screen) carrying the DNA Southern blots were hybridized as described by Silhavy et al. (12). The last posthybridization wash was carried out with 0.5 X SSC instead of 0.1 X SSC. Hybridization was detected by autoradiography (Fuji film RX) at -80°C with intensifying screens.

Results and discussion

The methodology and results of the bacterial matings are summarized in table II.

Transposon Tn5 confers neomycin and kanamycin resistance and was chosen for these studies because the Azospirillum strains we study are naturally sensitive to low levels of kanamycin (2-5 μg/ml). Moreover it was shown

Table II

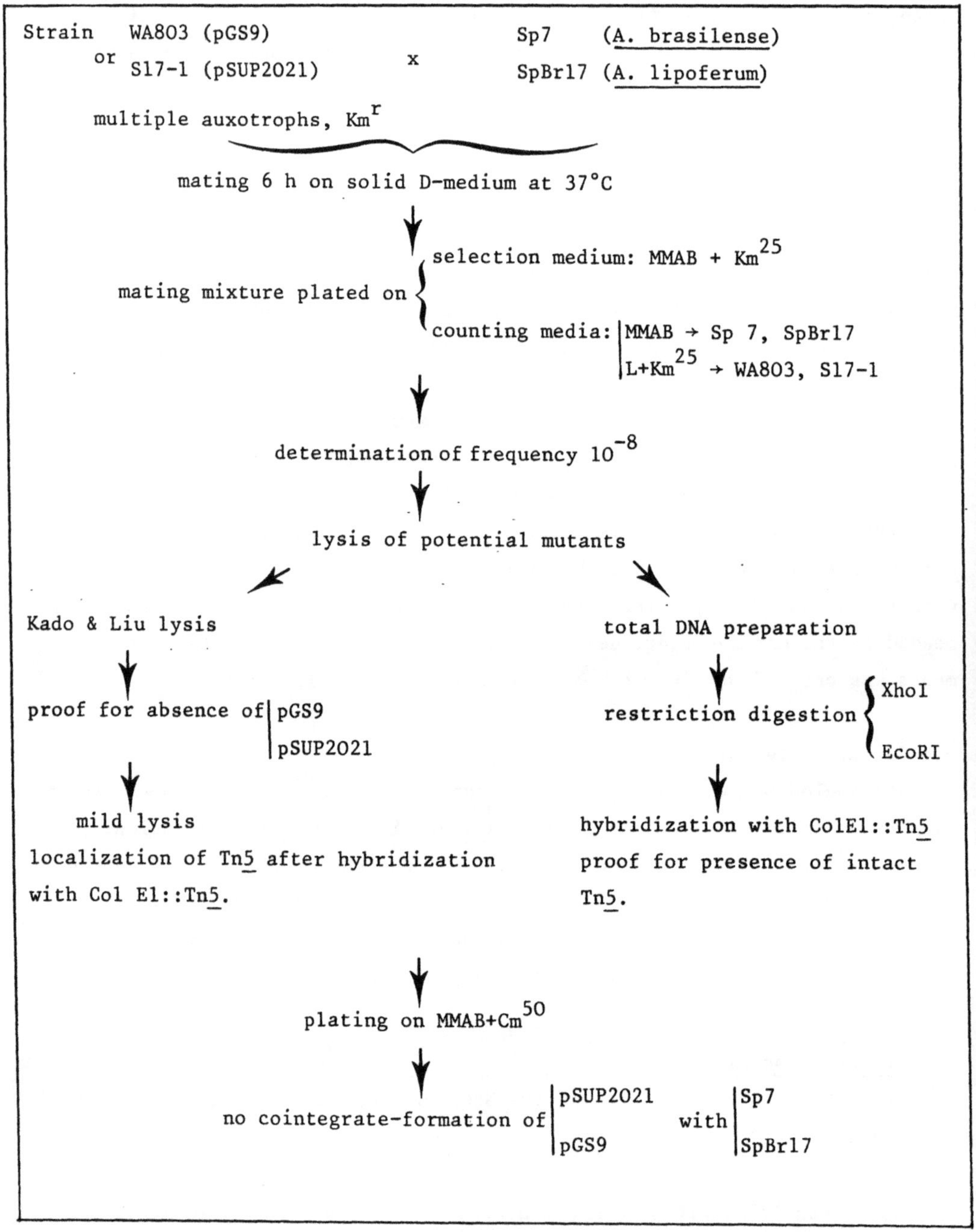
Strain WA803 (pGS9)
or S17-1 (pSUP2021)
x
Sp7 (A. brasilense)
SpBr17 (A. lipoferum)
multiple auxotrophs, Km^r
mating 6 h on solid D-medium at 37°C
mating mixture plated on
selection medium: MMAB + Km^25
counting media: MMAB → Sp 7, SpBr17
L+Km^25 → WA803, S17-1
determination of frequency 10^-8
lysis of potential mutants
Kado & Liu lysis
proof for absence of pGS9
pSUP2021
mild lysis
localization of Tn5 after hybridization with Col E1::Tn5.
total DNA preparation
restriction digestion
XhoI
EcoRI
hybridization with ColE1::Tn5
proof for presence of intact Tn5.
plating on MMAB+Cm^50
no cointegrate-formation of pSUP2021
pGS9
with Sp7
SpBr17

that Tn5 causes little site specificity and causes polar mutations (13). Previously it was already shown that Tn5 confers kanamycin resistance to Azospirillum (4).

To obtain Tn5 transpositions to Azospirillum replicons (chromosome and indigenous plasmids) we adopted two independent suicide plasmid techniques, described by Simon et al. (5) and by Selvaraj and Iyer (6). In our first attempts using the conditions as outlined in Table II, kanamycin resistant colonies of A. brasilense Sp7 and A. lipoferum SpBr17 were obtained at rather low frequencies (10^{-8} per recipient cell). However it is important to notice that by the way of selecting putative exconjugants (minimal medium plus kanamycin), auxotrophic mutants induced by Tn5 insertions are not detected. Since the first objective of this study is directed towards the physical analysis of presumptive Tn5 transposition strains, that is to detect kanamycin resistance in Azospirillum as the result of a simple at random Tn5 transposition event, no further attempts were made so far to obtain more accurate data on frequencies of Km^r exconjugants. Instead nineteen Km^r colonies of Azospirillum, picked at random from the four mating experiments as outlined in Table II, were used for further study.

Since the two suicide plasmids used, pSUP2021 and pGS9, also code for chloramphenicol resistance, Km^r exconjugants were tested on MMAB medium containing chloramphenicol (50 µg/ml). None of the colonies tested were able to grow on this medium, even after prolonged incubation. More substantial evidence for the absence of the suicide plasmids in the Km^r exconjugants is given by the plasmid detection method of Kado and Liu (10). Figs 1 and 2 demonstrate that the Km^r exconjugants have not acquired a new plasmid, following conjugation with E. coli donor cells. Therefore we predicted that the kanamycin resistance of the selected Azospirillum mutants is due to a simple Tn5 transposition event to new locations on the Azospirillum genome. To confirm the presence of Tn5 sequences in the mutants, Southern blots of Eco RI or Xho I digests of total DNA from eight of the mutants were probed with ^{32}P-labeled ColE1::Tn5 DNA (Figs. 3 and 4). Because Tn5 (5.7 kbp) contains no recognition site for Eco RI, an Eco RI digest of total cellular DNA from a strain carrying a single Tn5 element should contain a single Eco RI fragment that will hybridize with a Tn5 probe. Fig. 3B shows an autoradiogram of a Southern filter blot of Eco RI digested total DNA from eight Km^r exconju-

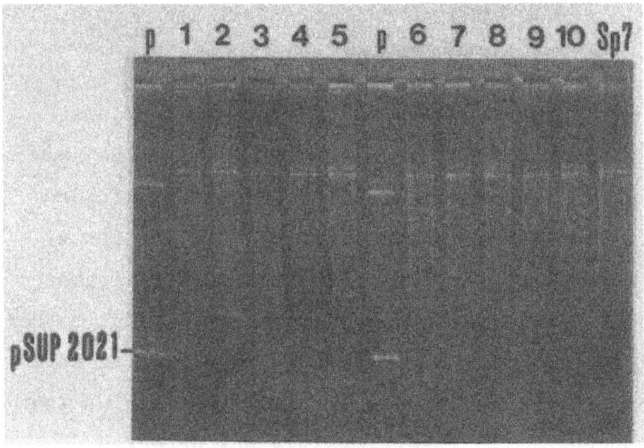

Fig. 1. Crude lysates (Kado and Liu, 10) of Km^r exconjugants of Sp7 (lane 1 to 10) generated by the system of Simon et al. (5) proving the absence of pSUP2021. p : S17-1 (pSUP2021)

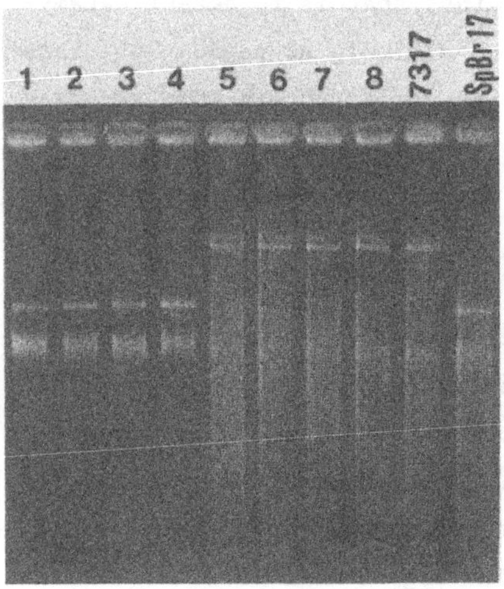

Fig. 2. Crude lysates (Kado and Liu, 10) of Km^r exconjugants of SpBr17 (lane 1 to 4) and Sp7 (lane 5 to 8) generated by the system of Selveraj and Iyer (6)

gants hybridized with ^{32}P-labeled Col El::Tn5 DNA. Except for mutant number 5, Tn5 sequences hybridize with a single fragment, indicating that these mutants carry a single Tn5 insertion. The result obtained with mutant number 5 can be explained either as the result of a partial digestion of the total DNA, or as the result of a multiple Tn5 insertion. Evidence in favor of the first possibility is given below.

To confirm the results obtained with Eco RI, total DNA from the same Tn5 transposition strains shown in Fig. 3 were cleaved with Xho I. Xho I clea-

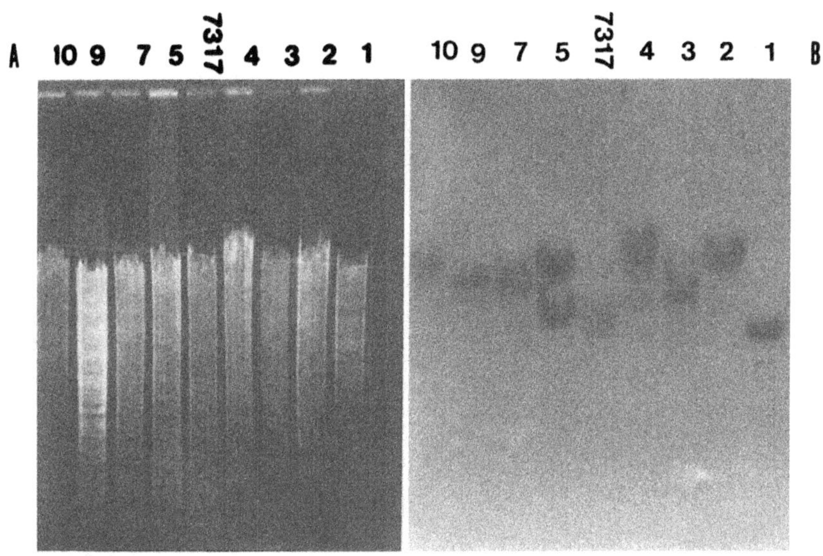

Fig. 3. A. EcoRI restriction digest of total DNA from Km^r exconjugants of Sp7 (lanes 1 to 10) (Simon et al., 5)

B. Hybridization with ^{32}P ColEI::Tn5 of the digest from Fig. 3 A. Azospirillum brasilense 7317, obtained by site directed Tn5 mutagenesis of the nifD-gene (Elmerich, 1984) was used as a positive hybridization control

ves Tn5 three times. This results in four fragments containing Tn5 sequences. Two are internal to Tn5 and will be the same wherever Tn5 is inserted. Two other are boundary fragments that will vary in size, depending on the size of the particular Xho I fragment of the Azospirillum genome into which Tn5 is inserted and are likely to be different for each independent Tn5 insertion. Fig. 4 clearly shows that Tn5 is inserted at different locations in the eight different strains. There is also clear evidence that mutant number 5 contains a single Tn5 insertion as predicted above. In order to locate the Tn5 insertions on the Azospirillum genome (chromosome or indigenous plasmids), crude lysates of the Km^r exconjugants were separated on an agarose gel, allowing the detection of plasmid DNA and chromosomal (linearized) DNA, as shown in Fig. 5A. Hybridization of Southern filter blots of such an agarose gel with ^{32}P-labeled Col El::Tn5 (Fig. 5B), allowed us to determine that of all Tn5 insertion mutants so far studied (total number of nineteen), Tn5 is integrated in the bacterial chromosome.

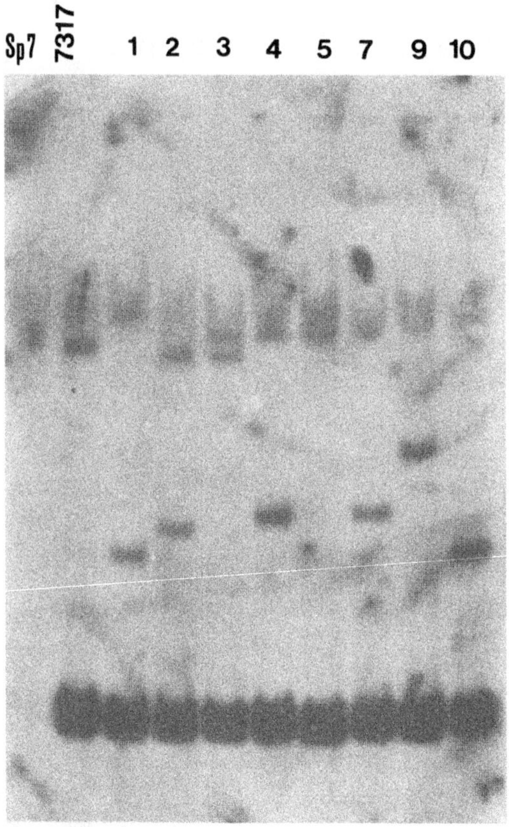

Fig. 4

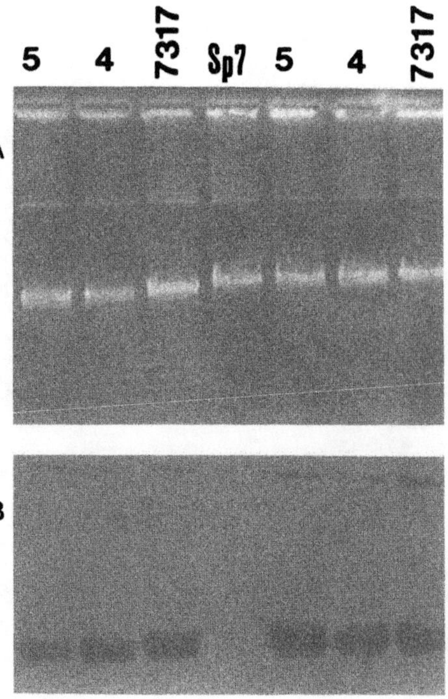

Fig. 5

Fig. 4. Hybridization with ^{32}P ColEI::Tn5 of the XhoI restriction digest of total DNA from Km^r exconjugants of Sp7 (lane 1 to 10). 7317: see Fig. 3 B. The two internal XhoI restriction fragments of Tn5 are indicated by arrows, the two boundary fragments are randomly distributed

Fig. 5. A. Crude lysates of two Km^r exconjugants of Sp7 (10). 7317: see Fig. 3 B

B. Hybridization with ^{32}P ColEI::Tn5 of the two lysates from A together with a positive (7317) and negative (Sp7) control. In both cases Tn5 insertion occured in the bacterial chromosome

In conclusion we have demonstrated that the two methods for transposon mutagenesis studied above, allow the isolation of Tn5 insertion mutants of Azospirillum. Our second objective is to determine and to optimize the frequency of transposition in Azospirillum. At present efforts are being made to enhance this frequency by studying conjugation and transposition under different conditions. Recent experiments (data not shown) indicate that by increasing the conjugation time between E. coli and A. brasilense Sp7 up to two days, Km^r exconjugants are obtained with a frequency of 10^{-6} (not counting the auxotrophic mutants).

Finally we are confident that by studying and optimizing other parameters involved in the conjugation and transposition events, transposon mutagenesis will become a powerful tool for the genetic analysis of chromosomal and plasmid encoded functions of Azospirillum spp.

Acknowledgements

We are grateful to Dr. Döbereiner (Brasil), Dr. Elmerich (France), Dr. Klingmüller (Germany) and Dr. Vlassak (Belgium) for providing us with Azospirillum strains.
This research was supported by grants from the "Fonds voor Kollectief Fundamenteel Onderzoek" (F.K.F.O. 2.0013.85) and from the PREST Program (KUL 44). M.V. is supported by a fellowship of the PREST Program, and K.M. is aspirant of the N.F.W.O.
"Deze tekst is opgesteld in het kader van het Programma ter versterking van het wetenschappelijk potentieel in de nieuwe technologieën - PREST (Diensten van de Eerste Minister - Programmatie van het Wetenschapsbeleid). De wetenschappelijke verantwoordelijkheid ervan berust bij de auteurs."

References

1. Döbereiner, J. and Day, J.M. (1976). In: Symposium on Nitrogen Fixation. Edited by W.E. Newton and C.J. Nyman. Washington State University Press, Pullman, pp. 518-538.

2. Elmerich, C. (1984). BIO/TECHNOLOGY 2, 967-978.

3. Singh, M. (1983). In: Klingmüller, W. (ed.). Azospirillum Genetics Physiology Ecology. EXS 48 pp 47-55, Birkhäuser, Basel.

4. Elmerich, C. and Franche, C. (1982). In: Klingmüller, W. (ed.). Azospirillum Genetics Physiology Ecology EXS 42 pp 44-52, Birkhaüser, Basel.

5. Simon, R., Priefer, U. and Pühler, A. (1983). BIO/TECHNOLOGY, nov., 184-190.

6. Selvaraj, G. and Iyer, V.N. (1983). J. Bacteriol. 156, 1292-1300.

7. Wood, W.B. (1966). J. Mol. Biol. 16, 118-133.

8. Franche, C. and Elmerich, C. (1981). Annales de microbiologie 132, n°1, 3-18.

9. In "Cours de microbiologie générale. Etude de la fixation de l'azote chez diverses bacteries". Elmerich, Institut Pasteur (1983).

10. Kado, C.I. and Liu, S.T. (1981). J. Bacteriol. 145, 1365-1373.

11. EMBO Course on "The use of Ti plasmids as cloning vectors for genetic engineering in plants". 4-23 August 1982. Lab. genetics, Rijksuniversiteit Gent, Belgium.

12. Silhavy, T.J., Berman, M.L. and Enquist, L.W. (1984). In: "Experiments with gene fusions". CSH Laboratory 1984.

13. Shaw, K.J. and Berg, C.M. (1979). Genetics 92, 741-747.

MUTATION IN AZOSPIRILLUM BRASILENSE

M. Maddalena Del Gallo, Loretta Gratani and G. Morpurgo
Department of Plant Biology, University of Rome
P.le Aldo Moro 5, I-00184 Rome, Italy

Summary: In order to study mutation frequencies in *A. brasilense* two selective systems, i. e. resistance to 5-Fluorotryptophan and to Rifampicin, were set up. The rate of mutation in this organism after treatments with several mutagens has been investigated. UV rays, X-rays, Diepoxybutane, 4-nitroquinoline-N-oxide and 8-methoxypsoralen + UV-A were almost completely ineffective in raising the mutation frequencies over the spontaneous one. UV rays were tested in a variety of physiological conditions always with negative results. Treatments with UV rays + Acriflavine and UV rays + 5-Bromouracil were negative as well. Ethyl methanesulfonate and Nitroseguanidine were the only mutagens tested which gave clearly positive results. On the base of these results the conclusion that *A. brasilense* is an SOS^- species is forwarded. Keywords: mutagen action, SOS repair system, photoreactivation, pR46.

Introduction

Free-living nitrogen-fixing Azospirilla have attracted considerable interest because of their ability to establish associations with the roots of grain crops: maize, wheat, sorghum and other Graminaceae (1,2,3). Extensive programs with the goal of improving this association and obtaining a consistent reduction of nitrogen fertilizers have been undertaken all over the world (4,5,6,7).

One of the main problems in the development of experimental programs has been the unforeseen difficulty in obtain stable mutations. Wood *et al.*(8) concluded that the difficulty met in mutagenizing *Azospirillum*, also after treatment with powerful mutagens, is due to the peculiar structure of its genome. According to these authors the *Azospirillum* genome does not have a main chromosome, but is composed of a set of different overlapping plasmids. The presence of multiple copies of each gene may explain the difficulty in obtaining mutations. These conclusions however are not confirmed by other authors (9,10) and

Azospirillum III: Genetics, Physiology, Ecology
Ed. by W. Klingmüller

the hypotesis seems to be improbable because of its genome size, which is only 1.8 times the genome of the E. coli.

The research we present here was undertaken to understand the difficulty met in obtaining mutants in A. brasilense, and to develop an efficient system for inducing mutations. To achieve these goals, two selective systems, i. e. resistance to 5-Fluorotryptophan and to Rifampicin, were devised and a variety of mutagenic treatments were used.

Material and Methods

Media and growth conditions

A. brasilense Cd (ATCC 29710) was grown aerobically in batch culture at 35°C in a minimal medium (MM), (11), supplementeded with one of the following carbon sources: Malic acid (neutralized with NaOH) 7 mM, Fructose 27 mM, Ethanol 1%.

Chemicals

DL-5-Fluorotryptophan (FT) was purchased from NBCO (Cleveland, USA); Rifampicin (3-4 methylpiperazinyl-iminomethyl rifamycin SV), Ethyl methanesulfonate (methanesulfonic Acid Ethyl Ester, EMS), Nitroseguanidine (1-methyl 3-nitro 1-nitroseguanidine, NTG), Acriflavine (a mixture of 2,8 (3,6) Diamino 10 methylacridinium chloride and 2,8 (3,6) Diaminoacridine), 4-Nitroquinoline-N-oxide (NQO) were purchased from Sigma (St.Louis, USA). Diepoxybutane (1,3 Butadiendiepoxide, DEB) was from Fluca (Buchs, Switzerland), 5-Fluorouracile (2,4-Dihydroxy 5-Fluoropyrimidin, 5-FU), 5-Bromouracil (5-BU) from Sckuchardt (München, FRG), Methoxypsoralen (8-methoxypsoralen, 8-MOP) was a gift from Franco Bordin, University of Padua.

Detection of the mutants

FT-resistant mutants were selected in Petri dishes seeding the bacteria on the surface of agarized MM medium enriched with 40 µg/ml of FT. This drug concentration is toxic to wilde type cells and the FT-resistant mutants are easily detectable. These mutants overproduce tryptophan. When tryptophan is secreted into the dish, it competes with the drug, thus detoxifyng the medium.

Each mutant is therefore surrounded by a growth-halo of wild type cells. Rifampicin mutants were selected adding 10 μg/ml of this compound to the agarized medium. In this case there is no residual growth on the dishes.

Mutagenic treatments

The following mutagenic treatments were performed on cells grown at the late exponential phase : DEB, X-rays, 8-MOP + UV-A, 4-NQO, EMS, NTG, UV.

UV rays were also used in conjunction with Acriflavine and with 5-BU. Cells were grown overnight before UV irradiation with Acriflavine to brake plasmid multiplication; instead, they were pre-treated with 5-BU to hamper the DNA repair process (12).

Moreover the UV rays, X-rays and DEB action were tested in cells grown under different physiological conditions.

In addition, cells were exposed to severe stress after UV-irradiation, according to the procedure described by Little and Mount (13).

Results

Diepoxybutane

Diepoxybutane (DEB) is a bifunctional alkylating agent and a powerful mutagen in most organisms because of its double epoxidic structure (14).

Cells were treated with DEB (0.2 mM) for 10 and 20 min .The mutants were detected on Rifampicin, FT and 5-FU enriched MM after two to five days of growth at 35°C. Results are shown in figure 1.

X-rays

X-rays are the first mutagenic agent discovered. They act on the genetic material with several direct and indirect system, inducing not only point mutations but every other possible type of genetic damage.

A Siemens Stabily Pan with a 0.2 mm Cupper filter has been used for these experiments. Exponentially growing cells were suspended in 0.9% NaCl solution

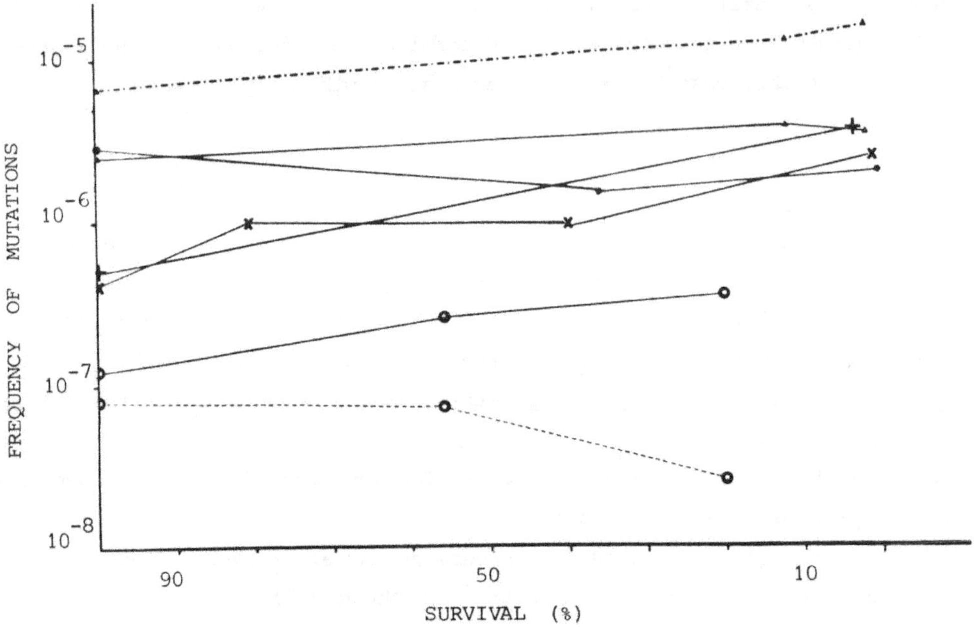

Fig. 1. Mutation frequencies of resistance to FT (——), FU (- - -) and Rifampicin (---) after treatments with 0.2 mM DEB. Each different experiment is marked by a different symbol.

and irradiated respectively with 2000, 4000, 8000, 12000 rad. Results are given in figure 2.

8-methoxypsoralen + UV-A (365 nm)

8-MOP is a compound which, when photoactivated by UV-A light, is a powerful mutagen producing monoadducts on the pyrimidine and extensive DNA crosslinking in several organisms (15). In the dark,however, this product is not active, and UV-A alone has generally no effect.

Cells were treated one hour with 10 µg/ml 8-MOP and then exposed 20 min to UV-A (Sylvania lamp,GTE USA, Blacklight Blue F 15T8/BLB 15 W). Previous experiments have demonstrated that this dose severely affects survival. Data given in table 1 show that UV-A + 8-MOP affects survival without altering the mutation frequencies.

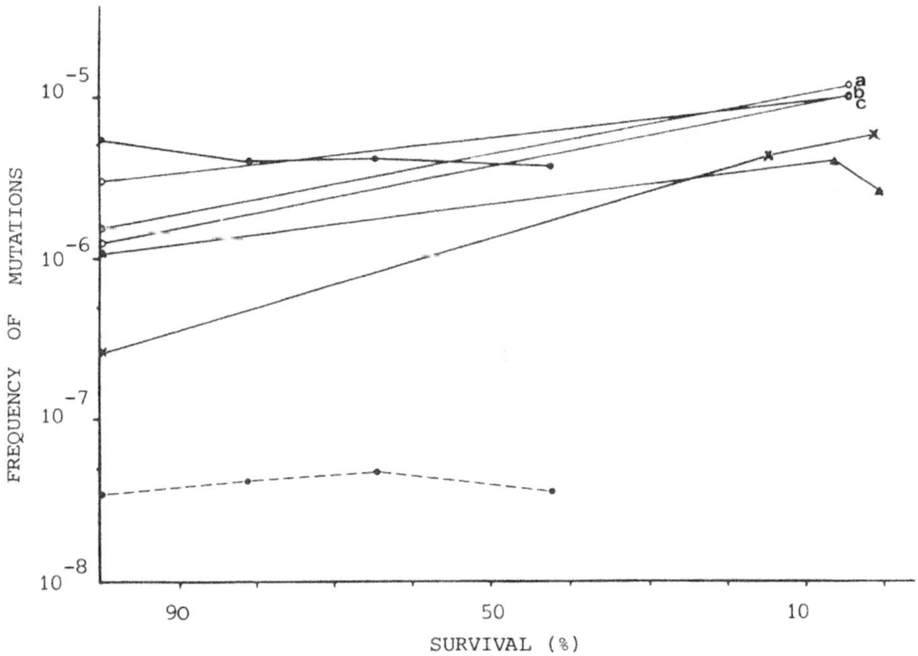

Fig. 2. Mutation frequencies to 5-FT (——) and to Rifampicin (---) after treatment with X-rays. **a** = Cells grown aerated up to the exponential phase; **b** =cells grown up to the exponential phase without aeration; **c** = cells grown aerated up to the stationary phase.

Table 1. Treatment with 8-MOP + UV-A

	Survival %	FT mutation frequency $x10^{-6}$	FT resistant colonies
Control	100	2.4	96
8-MOP + UV-A	11.8	2.6	187

4-nitroquinoline-N-oxide

NQO is known to be a powerful mutagen that acts similarly to the UV light. It induces single mutations and it is presumably specific to GC base pairs (16).In Azospirillum this compound does not affect survival neither mutation frequency.

UV rays treatments

Exponentially growing cells were treated on the surface of Petri dishes in the daylight for 20 and 40 sec. An Osram 15 W (253.7 nm) lamp,at a distance of 44 cm was used. Mutants were detected on Rifampicin and FT enriched MM after 2 to 5 days of growth at 35°C. Results are shown in figure 3.

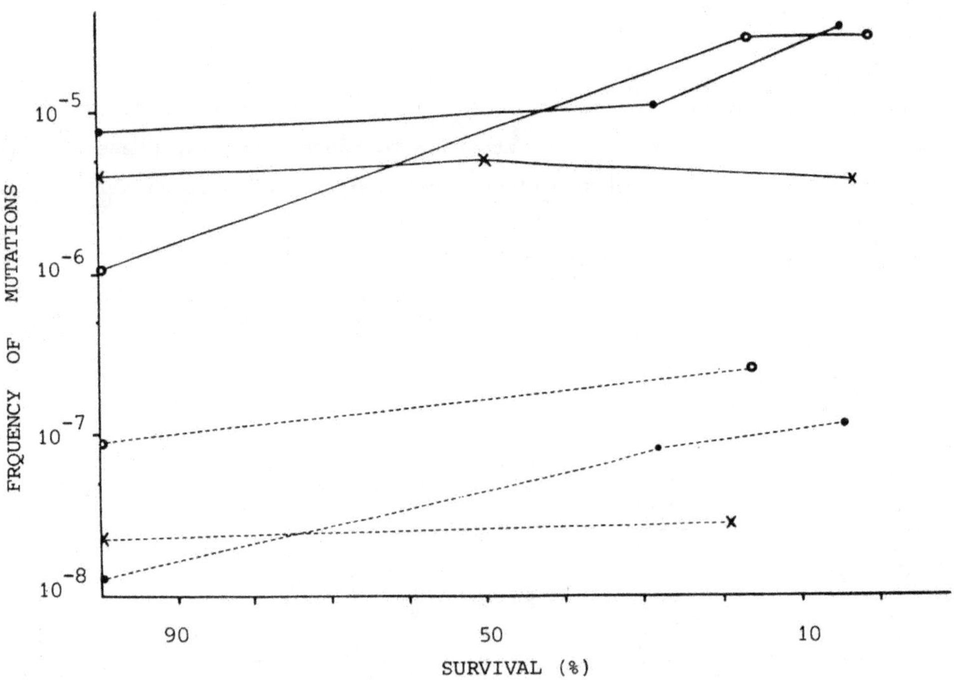

Fig. 3. Mutation frequencies of resistance to FT (——) and to Rifampicin(---) after UV treatments.

In order to ascertain if the failure in obtaining consistent increase in the mutation frequency after UV treatment was due to a photoreactivation process, an experiment was carried out with irradiation in daylight and in the dark. At the same time the influence of certain physiological conditions was tested, that is, whether the differents poli-β-hydroxybutyrate (PHB) contents of the cells influence the survival and mutation frequency of Azospirillum after irradiation.

The bacteria were grown in MM supplemented with fructose or with ethanol as carbon sources up to the beginning of the stationary phase; they were seeded and irradiated with UV in light or in the dark and successively incubated in the same conditions. PHB content was tested by the Ward and Dawes disk assay method (17) before irradiation.

Cells cultivated with fructose have a higher content of PHB (60 %) than cells cultivated with ethanol (5 %). This is probably due to energetic difficulties in synthesizing reserve material in these particular conditions. The results are given in figure 4.

These results are in agreement with Okon's findings (18) and seem to show that in Azospirillum an intense photoreactivation process is active. Besides, other different mechanisms of protection from the lethal action of UV rays. correlated with the amount of PHB, should be present. In figure 4 the major resistance to UV of PHB-rich cells as compared to the PHB-poor cells is evident. Frequency of mutation is however lightly influenced by photoreactivation.

Treatment with UV rays + Acridine and with UV rays + 5-BU

In order to lower the number of plasmids in Azospirillum, cells were treated with Acridine before UV irradiation. It is known, in fact, that Acridine inhibits selectively the multiplication of plasmids (19).

Bacteria were grown overnight in Nutrient broth (Difco, Detroit, USA) to which 2.5 µg/ml of Acridine was added. This is the maximum concentration that allows cells to multiply. The bacteria were then irradiated for 40 sec with UV rays on the surface of the dishes. Results are given in table 2.

It is known that DNA which has incorporated 5-BU is not repairable by photo or dark repair processes (15).

Bacteria were grown in four successive transferts in MM supplemented with thimine (50 µg/ml) to brake its endogenous synthesis. The cells were then washed twice and resuspended in MM + 5-BU (50 µg/ml), incubated 5 hrs and treated with UV rays (Table 2).

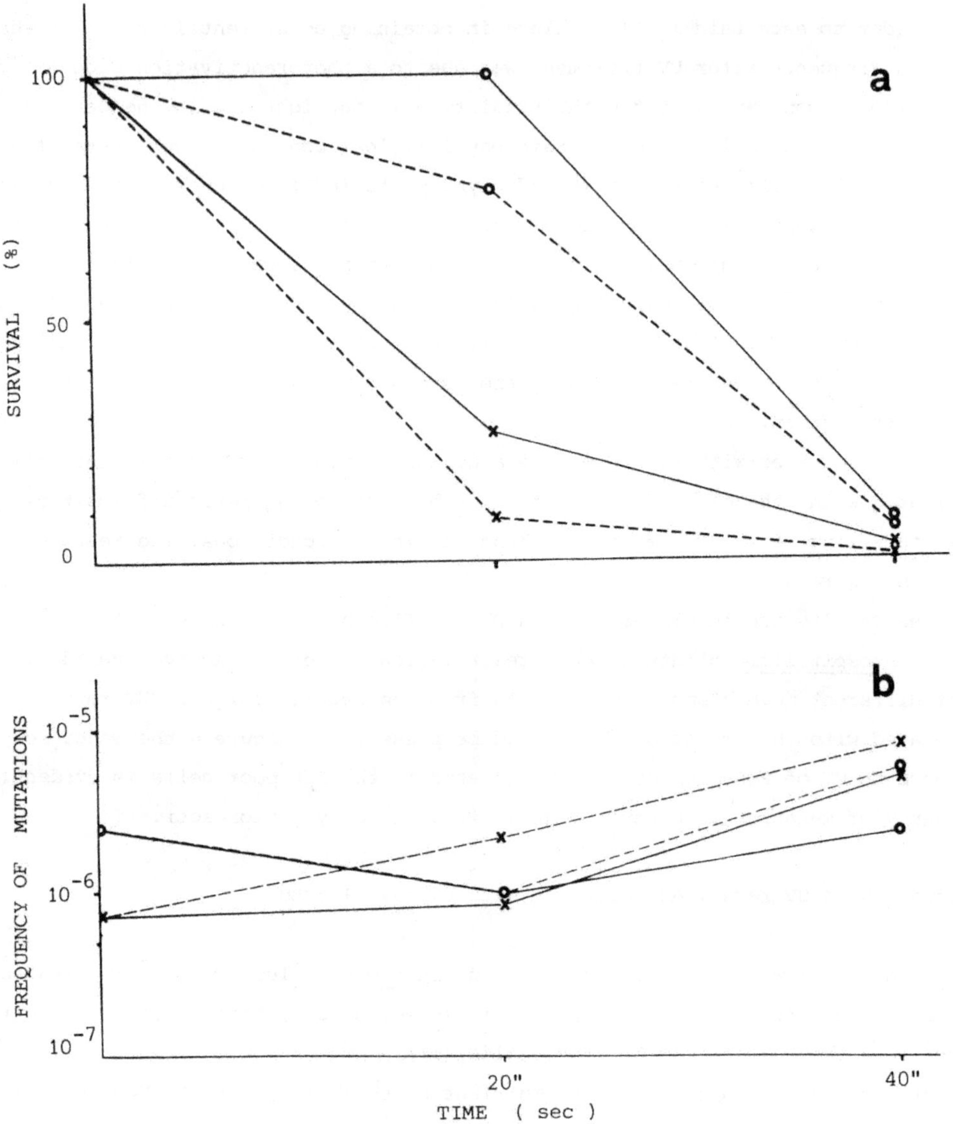

Fig. 4. **a** = Survival of cells after UV treatment in the dark (----) and in the light (——). Fructose grown cells (x——x) and ethanol grown cells (o——o). **b** = Mutation frequencies (symbols as in a).

Both treatments - UV rays + Acridine and UV rays + 5-BU - produce a moderate increase in the frequency of mutants on the survivors, but the increase is only relative while there is a decrease in the absolute number of mutants on the plated cells.

UV treatment associated with physiological stress

The failure in obtaining mutations with UV rays and DEB utilizing also, as described above, cells grown in different physiological conditions, may be explained as a lack of SOS repair system in this bacterium. We therefore tried a complex treatment (13) which in E. coli induces the SOS system at the maximum level.

Exponentially growing cells were suspended in 0.03 M pH 7 phosphate buffer to a density of 10^9 cell/ml. A thin layer of suspension was stirred during irradiation in an open Petri dish (20 ml of suspension in a 15 cm glass dish). The UV dose was adjusted to give a 10^{-3} survival rate. After irradiation the cell suspension was kept at 4°C in the refrigerator for 24 hrs to stabilize DNA damage, and finally resuspended in MM without a carbon source and incubated for several periods at 20°C. Cells were then plated after 15 to 60 days and the frequency of FT-resistant mutants was scored. No significant increase in mutation frequencies was detected. The whole experiment was performed in the dark to prevent the photorepair process.

It has been recently described that in a different family of nitrogen-fixing bacteria, the Rhizobiaceae, which are depleted of the SOS repair system, this treatment leads to the unexpected transformation of the original bacterium in five different species (20). We have therefore examined this possibility also in Azospirillum. The only significant difference in the colonies recovered after the stress, was only a higher utilization of glucose which is normally a very poor carbon source for A. brasilense. Nevertheless the same physiological change occurs also in cells exposed to stress condition without UV irradiation.

Ethyl methanesulfonate

EMS is an alkylating agent that acts on DNA primarily by a modification of guanine (O^6-alkyl guanine), (21).

Table 2. Mutation frequency of resistance to 5-FT after treatments with UV.
a) Cells grown overnight in MM+Acridine
b) Cells grown 5 hrs in MM + 5-BU

	Time sec UV	Survival %	Mutation frequency $x10^{-6}$	FT-resistant colonies
a)	0	100	1	49
	40	1.9	13.7	123
b)	0	100	2.6	148
	40	0.9	26	133

A bacterium culture,grown overnight in MM up to the late logarithmic phase, was diluted in equal volume with a mixture of fresh MM + 0.04 ml/ml of EMS and treated for 30 and 60 min at 30°C. Results are given in figure 5.

Nitroseguanidine

The NTG is one of the most powerful mutagens yet discovered for bacteria. It induces primarily base transition mutations of the GC-AT type although AT-GC transitions, transversions and even frameshifts arise at low frequencies. It induces lesions that require SOS processing which in turn induce mutations. NTG can act, like EMS, as mutagen by inducing lesions such O^6-methylguanine that cause mutation by directly mispairing, bypassing the SOS repair system. Overnight grown culture (MM) was treated with 50 and with 100 ug/ml of NTG.

The results obtained - reported in figure 6 - similarly to those obtained with EMS treatments, show that both mutagens are able to increase the mutation frequency in *Azospirillum*

Discussion

The data reported show that *Azospirillum brasilense* is almost completely insensitive to mutagenic action of a large variety of treatments which are very powerful on most of the organisms. Any attempt to induce mutations with UV rays

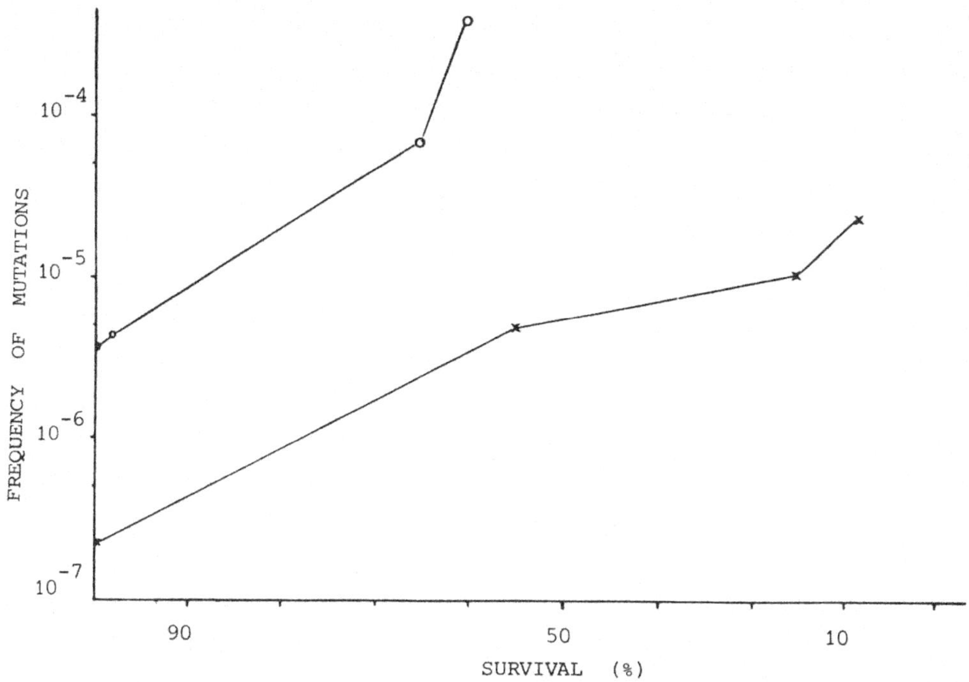

Fig. 5. Mutation frequencies of resistance to FT after treatments with EMS o=30', x= 60'

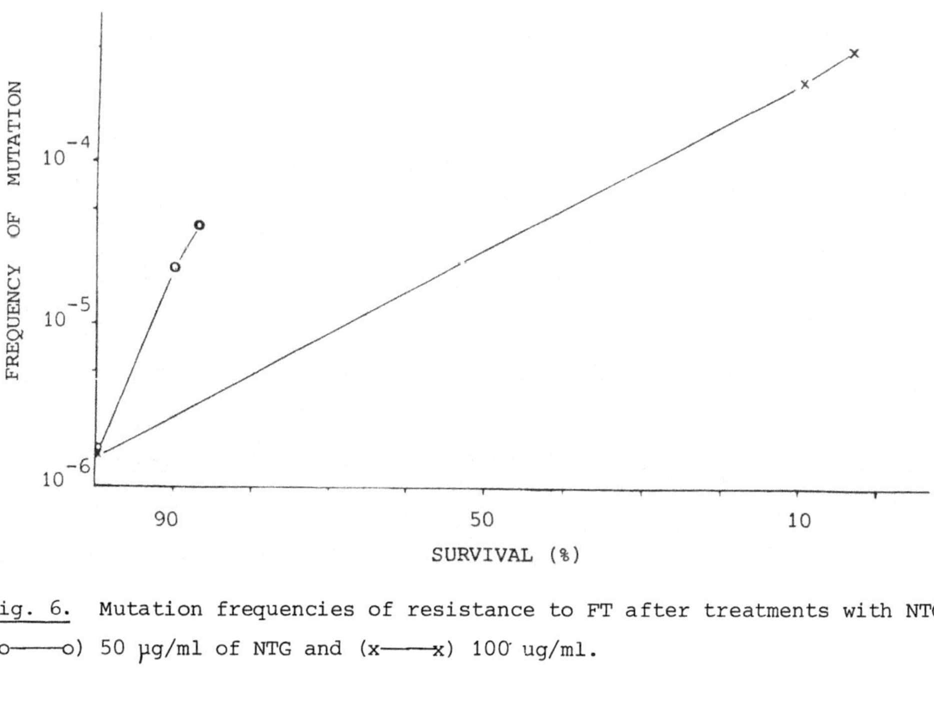

Fig. 6. Mutation frequencies of resistance to FT after treatments with NTG. (o——o) 50 μg/ml of NTG and (x——x) 100 ug/ml.

changing also the physiological conditions and stressing the cells, was unsuccessful. In some experiments the frequency of mutation on surviving was higher than the control. However, we have never found an absolute increase in the number of the mutants. The apparent increase may be actually due to selection of the preexistent clones of mutants.

Only NTG and EMS, between the mutagens tested, were effective in raising mutation frequencies beyond the spontaneous one. The most likely explanation of the unusual behavior of A. brasilense is that this bacterium, like some other (Hemophilous influenzae, Proteus mirabilis etc., 22), lacks SOS functions.

This explanation is in agreement with the fact that the only mutagens which clearly increase the mutation frequencies beyond the spontaneous ones are EMS and NTG. These mutagens may induce mutations without the intervention of the SOS repair system (22). In absence of SOS repair many DNA-damaging agents are lethal for the cells, but their mutagenic action is nil or very weak.

The attempt to demonstrate this hypothesis also by introducing in A. brasilense the plasmid R46 bearing part of the SOS repair system from E. coli was unsuccessful because this plasmid, selected for tetracycline resistance, did not establish in Azospirillum. Positive results obtained by using NTG and EMS mutagenic agents, also in absence of an SOS repair system are in agreement with this explanation.

References

1. Döbereiner J., J.M. Day (1976). In "Proc. 1st. Int. Symp. N_2-Fixation". W.E. Newton, C.J. Nyman (Eds). Wash. State Univ. Press, Pulmann, 518-538.

2. Smith R.L., J.H. Bouton, S.C. Schank, K.H. Quesenberry, M.E. Tyler, J.R. Milam, M.H. Gaskins, R.C. Littell (1976). Science 193, 1003-1005.

3. Albrecht S.L., Y. Okon, J. Lonquist, R.H. Burris (1981). Crop Science 21, 301-306.

4. Vlassak K., L. Reynders (1981). In "Associative N_2- Fixation". P. B. Vose, A.P. Ruschel (Eds). CRC Press Inc. Boca Raton, Fla. 1, 93-101.

5. Albrecht S.L., M.H. Gaskins, J.R. Milam, S.C. Schank, R.L. Smith (1983). In "Azospirillum II" W. Klingmüller (Eds). EXS48, Birkhäuser, Basel, 138-148.

6. Hegazi N.A., M. Monib, H.A. Amer, E.S. Shokr (1983). Can. J. Microbiol. 29, 888-894.

7. Okon Y. (1984). In "Advances in Nitrogen Fixation Research". C. Veeger, W.E. Newton (Eds). Nijhoff - Junk, Pudoc., 303-309.

8. Wood A.G., E.M. Menezes, C. Dykstra, D.E. Duggan (1982). In "Azospirillum". W. Klingmüller (Eds). EXS42. Birkhäuser, Basel, 18-34.

9. Franche C., E. Canelo, D. Gauthier, C. Elmerich (1981). Microbiol. Lett. 10, 199-202.

10. Plazinski J., P.J. Dart, B.G. Rolfe (1983). J. Bacteriol. 155, 1429-1433.

11. Martinez-Drets G., M. Del Gallo, C. Burpee, R.H. Burris (1984). J. Bacteriol. 159, 80-85.

12. Setlow R.B., J.K. Setlow (1972). Ann. Rev. Bioph. Bioenerg. 1, 293.

13. Little J.W., D.W. Mount (1982). Cell 29, 11-22.

14. Morpurgo G. (1963). Genetics 48, 1259-1263.

15. Scott B.R., T. Aldersen (1971). Mutation Res. 12, 29-34.

16. Bal J., E.M. Kajtaniak, N.J. Pieniazek (1977). Mutation Res. 56, 153-156.

17. Ward A.C., E.A. Dawes (1973). Analytical Biochem. 52, 607-613.

18. Tal. S., Y. Okon. Can. J. Microbiol., in press.

19. Novick R.P. (1969). Bacteriol. Rev. 33, 210-263.

20. Heumann W., A. Rösch, R. Springer, E. Wagner, K.P. Winkler (1984). Mol. Gen. Genet. 197, 425-436.

21. Lindahl T. (1982). Annu. Rev. Biochem. 51, 61-87.

22. Walker G.C. (1984). Microbiol. Rev. 48, 60-93.

PRODUCTION OF INDOL-3-ACETIC ACID BY AZOSPIRILLUM BRASILENSE

S. Horemans and K. Vlassak
K.U.Leuven, Faculty of Agriculture
Laboratory of Soil Fertility and Soil Biology
Kard. Mercierlaan 92, B-3030 Leuven, Belgium

Summary : Indol-3-acetic acid (IAA) was analysed in the culture medium of *Azospirillum brasilense* using a fluorimetric detector coupled to a reversed phase high performance liquid chromatography column. Accuracy of the analysis was confirmed by successive approximation and additionally verified by the 2-methyl indolo-α-pyrone assay. When *A. brasilense* was grown aerobically and in the presence of ammonium up to 40 $ng.ml^{-1}$ IAA was excreted into the medium in the absence of any added tryptophane. IAA concentrations in the medium increased mainly during the early stationary phase and remained remarkably constant afterwards. Keywords : *Azospirillum*, plant growth substances, indol-3-acetic acid (IAA), analysis of indol-3-acetic acid.

Introduction

Much attention has been given to the effect of *Azospirillum spp.* inoculation on growth parameters of cereals. A commercially relevant increase in one of the yield components has been observed in a considerable number of studies from different laboratories (1,2,3,4,5). The mechanism of these bacterial effects remains far from clear. Among the authors quoted only Baldani and coworkers (1) could explain part of their results by the nitrogenase activity of the bacteria. By authors from all five laboratories (1,2,3,4,5) the possible

Azospirillum III: Genetics, Physiology, Ecology
Ed. by W. Klingmüller

production of plant growth substances by the bacteria has been given as an alternative hypothesis. Arguments in favour of this hypothesis are given by the fact that in some instances the inoculation effect could be minimiced by the application of one or more growth substances (6,7,8,9). Furthermore bacterial culture medium shows cytokinin-like activity in the chlorophyl retention assay and gibberellin-like activity in the lettuce hypocotyl elongation assay (6). However available information lacks any data on reproducibility, accuracy or authenticity of the response and can for this reason not be accepted untill further confirmation. It has been demonstrated by bio-assay (6,10) and colorimetry (11) that Azospirillum spp. can convert tryptophane into indol-3-acetic acid (IAA). Mutant strains produce large amounts of IAA in the absence of added tryptophane (11). In inoculation studies using these strains (19) differences between wild type and mutant strains were minimal. To our knowledge IAA synthesis in the absence of added tryptophane has not been demonstrated yet. To summarise, very few data do sustain the hypothesis that plant growth substances have a role in the association between Azospirillum and plants. This is partly due to the absence of a suitable system to assay in a reproducible manner this association or to evaluate potential beneficial effects of inoculation. Furthermore interactions between microorganisms and the hormone balance of the plant may be very complex (12) and estimating trace amounts of plant growth substances still is a serious technical problem (13). Finally, despite intensive research effort the role of plant growth substances in the life of the intact growing plant is far from clear (14). Against all odds we have initiated a research program on the role of plant growth substances in the association between cereals and Azospirillum. In this communication we will present and discuss some of the results on bacterial IAA production in vitro.

The analysis of auxin-like activity

Auxins are operationally defined as substances that promote elongation in shoot cells, primarily in detached shoot segments and the final decision whether a micro-organisms produces auxines or not, must be made on the activity in bio-assay. Nevertheless, since IAA still remains the dominant auxin of physiological significance, since its production has been demonstrated in other bacteria and since modern physico-chemical methods are more sensitive, accurate and reproducible than bio-assays we decided to verify the presence of IAA first and screen for other auxins or anti-auxins at the appropriate stages later on.

The usefulness of a physico-chemical analysis depends completely on its accuracy : one must quantify IAA and nothing but IAA or the effort is meaningless. Two possible strategies for verifying accuracy have been discussed in full detail by Reeve and Crozier (12). The first method requires that a probability level should be placed upon the number of compounds potentially present in the extract. The analysis is conducted in such a way that sufficient information is obtained from the quantified response to distinguish the hormone from all other possible compounds within the choosen probability levels. This method essentially attempts to reveal possible sources of outside interference. The other strategy (12) has been called successive approximation and is based on the fact that subsequent, not correlated chromatographic procedures reduce the complexity of the extract and ultimately eliminate sources of interference within definable limits of probability. We followed the latter approach.

Materials and methods

Azospirillum brasilense strain R07 was obtained from Dr. J. de Lafonteyne (Rijksuniversiteit Gent, Belgium). This strain was originally isolated by Rinaudo (16) from rice. $1cm^3$ of a 24 hr old culture was washed twice with phosphate buffer (pH 6.6, 0.01 M) and inoculated in 100 cm^3 erlenmeyer flasks containing 25 cm^3 of liquid medium (17) and 1 $g.dm^{-3}$ NH_4Cl. Yeast extract, biotin or bromothymol were omitted to avoid any misinterpretation of the analysis. For kinetical studies daily 1 cm^3 of the you g s culture was transferred into fresh medium and all experiments were done on the same day. The first and second transfer were not used.

The HPLC equipment consisted of an Altex 110A apparatus equipped with an Altex 210 injection system and a 20 mm^3 injection loop and connected with a Shimadzu RF530 fluorescence detector (excitation at 285 nm and emission at 360 nm, a 17 mm^3 cuvette). The column (25 x 4.6 cm I.D.) was packed with ROSIL C18 HL, 5 micrometer particle size and eluted isocratically at 1 $cm^3.min^{-1}$. The eluens for ion-suppression chromatography was methanol:distilled water:acetic acid (50:50:0.5 v/v) and for ion pairing chromatography methanol:0.01 M tetrabutylammonium hydroxide in 0.001 M phosphate buffer (pH 6.6)(45:55 v/v).

The 2-MIP assay and the prepurification procedure on C18 prepacked cartridges were done according to (18). Turbidity was measured at 530 nm.

Results

The supernatant of a 6-day old culture of A. brasilense strain R07 was purified on two different HPLC systems and analysed by on-line fluorimetry. Following as well ion-suppression reversed phase (IS) chromatography (figure 1) as ion-pairing (IP) chromatography (figure 2 a peak was found co-chromatographing with authentic IAA. This peak was not found when an autoclaved cell suspension had been inoculated in the same medium (data not shown). The absolute concentrations calculated were 39.2 ± 1.6 and 44.0 ± 2.5 $ng.ml^{-1}$ supernatant following IS and IP chromatography respectively (triplicate analysis). Using (figure 3) both chromatographic procedures in series (first IS and then IP chromatography) a concentration of 40.9 ± 3.6 $ng.ml^{-1}$ was measured at this final purification level. Differences between both estimates can be explained by random variation on the analysis and the value obtained after IS could be accepted as the accurate index.

In a separate experiment we compared the analysis by on-line fluorimetry following IS chromatography with an estimate obtained by the 2-methyl indolo-α-pyrone (2-MIP) assay on the collected IAA fraction. In order to obtain sufficient material to do the 2-MIP assay we had to concentrate a larger amount (40 ml) of supernatant. This had to be pre-purified before HPLC. This prepurification was done on C18 Sep Pak cartridges (see M&M). Since the separation mechanism on the C18 cartridges is identical to the one during the HPLC step we can argue that this highly concentrated sample is not further purified than the one in the first experiment reported. We measured 26 ± 1.2 $ng.ml^{-1}$ by on-line fluorimetry and 31.59 ± 14.38 $ng.ml^{-1}$ by the 2-MIP assay. Although the standard error on the 2-MIP assay was unusually large in this particular experiment, the estimate is in accord with the on-line fluorimetric analysis. Absolute values are about 30 % lower than in the first experiment which is most probably due to losses during the extra manipulations.

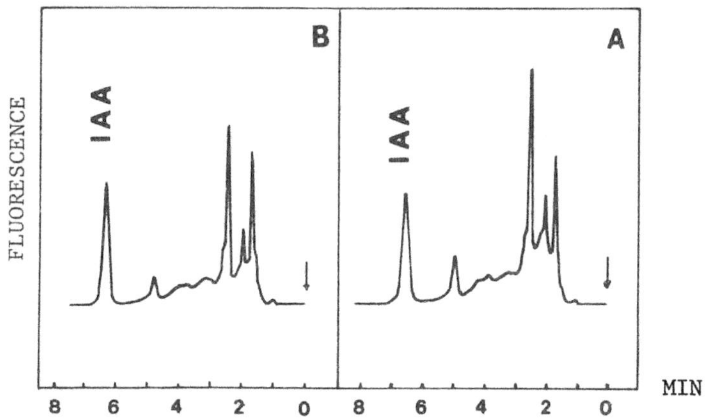

<u>Fig. 1</u> : Analysis of culture supernatant of <u>Azospirillum brasilense</u> (R07) by on-line fluorimetry after purification by ion-suppression reversed phase high performance liquid chromatography (see M&M). A : 20 μl supernatant; B : 15 μl supernatant + 1.75 ng IAA.

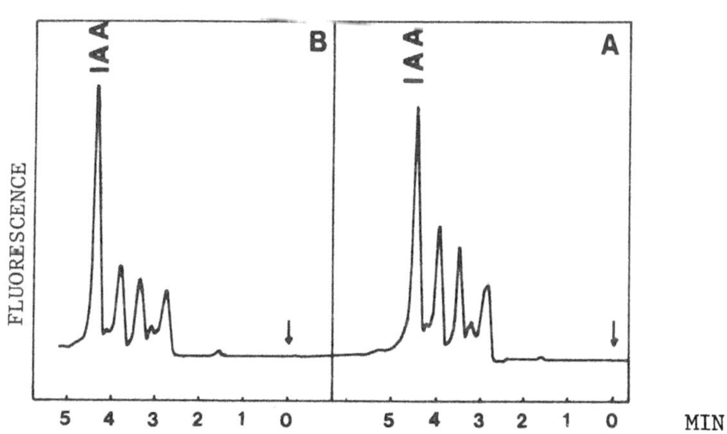

<u>Fig. 2</u> : Analysis of culture supernatant of <u>Azospirillum brasilense</u> (R07) by on-line fluorimetry after purification by ion-pairing high performance liquid chromatograph (see M&M). A : 20 μl supernatant; B : 15 μl supernatant + 1.75 ng IAA).

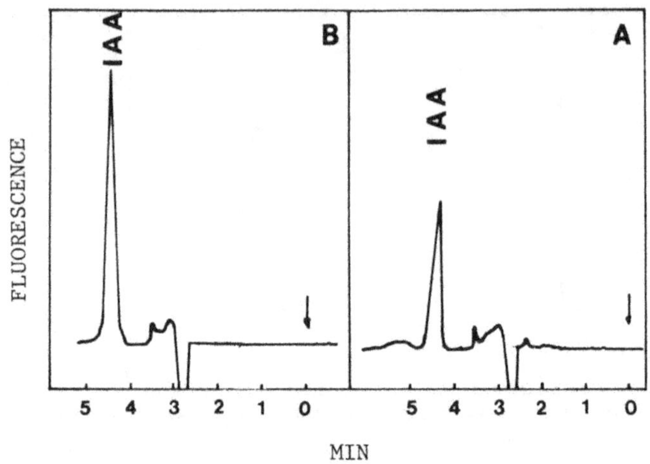

Fig. 3 : Analysis of culture supernatant of Azospirillum brasilense (R07) by on-line fluorimetry. The supernatant was first purified by ion suppression reversed phase chromatography and the fraction at the retention time of IAA was subjected to ion pairing chromatography (see M&M). A : extract only, B : extract + IAA.

In figure 4 the kinetics of both bacterial growth and IAA concentrations in the supernatant are given. Turbidity was maximal after 14 hours of growth and remained constant afterwards. IAA concentrations increased untill the third day after inoculation and remained constant afterwards.

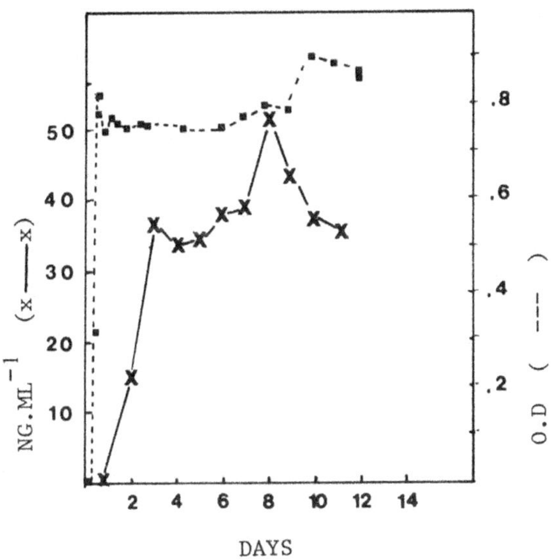

Fig. 4 : Bacterial growth (O.D. at 530 nm) and IAA concentration in the culture supernatant (ng.ml^{-1}) of *Azospirillum brasilense* (R07) growing aerobically in the presence of 1 g/l NH_4Cl.

Discussion

In the absence of added tryptophane up to 40 ng.ml^{-1} IAA was found in the culture medium of *A. brasilense* strain R07. Accuracy of the analysis was tested by successive approximation and further verified by the 2-MIP assay. The use of a spectrofluorimeter coupled to a reversed phase high performance column permits selective detection and accurate estimation of low levels of IAA in crude culture supernatant. The presented methodology offers a powerful tool for more extended studies on factors affecting IAA excretion in pure culture. Nevertheless when yeast extract, soil extract or plant exudates should be added to the medium or when other bacteria should be studied accuracy must be retested.

If this partner independent IAA synthesis occurs also in association with plants remains to be demonstrated.

The increase of IAA into the culture medium occured predominantly during the early stationary phase. This increase is not necessarely due to an active excretion of the hormone. It has been shown that passive diffusion of IAA through biological membranes occurs when IAA is in the non-dissociated form (15). The hormone (pK = 4.7) will distribute between cells and the surrounding medium according to a difference in pH between both compartments, being trapped in the more alkaline one. It is also known that *Azospirillum* causes an increase in the pH of the medium.

From three days after inoculation on IAA concentrations remain remarkably constant. Since IAA is readily oxidized this constant level most probably represents an equilibrium between destruction and synthesis of the hormone.

Acknowledgements

The authors gratefully acknowledge the financial support of the I.W.O.N.L. (grant no. 4383). They also wish to thank S. DEMARSIN for the skilful technical assistance and M.J. STRUYVEN for preparing the manuscript.

References

1. Baldani, V.L.D., Baldani, J.I. and Döbereiner, J. (1983). Can. J. Microbiol. 29, 924-929

2. Millet, E., Avivi, Y. and Feldman, M. (1984). Plant and Soil 80, 261-266

3. O'Hara, G.W., Davey, M.R. and Lucas, J.A. (1981). Can. J. Microbiol. 27, 871-877

4. Okon, Y., Heytler, P.G. and Hardy, R.F.W. (1983). Applied and Env. Microbiol. 46 (3), 694-697

5. Reynders, L. and VLASSAK, K. (1982). In "Azospirillum Genetics, Physiology, Ecology" Experientia Suppl., Klingmüller (Ed.) 42, 121-129

6. Tien, T.M., Gaskins, M.H. and Hubbell, D.H. (1979). Applied and Environm. Microbiol. 37 (5), 1016-1024

7. Inbal, E. and Feldman, M. (1982). Israël J. Botany 31, 257-263

8. Dubois, M. (1983). Ph.D. Thesis, Institut national agronomique, Paris

9. Plazinski, J. and Rolfe, B.G. (1985). Applied and Environmental Microbiol. 49 (4), 990-993

10. Reynders, L. and Vlassak, K. (1979). Soil Biol. Biochem. 11, 547-548

11. Hartmann, A., Mahavir Singh and Klingmüller, W. (1983). Canadian J. Microbiol. 29 (8), 916-923

12. Pegg, G.F. (1976).
In "Physiological Plant Pathology", Hertefuss and Williams (Eds.) 560-591

13. Crozier, A. and Reeve, D.R. (1980).
In "Hormonal regulation of development I", McMillan (Ed.), Enc. of Plant Physiology New Series 9, 203-280

14. Trewavas, A. (1981).
Plant, Cell and Environment 4, 203-228

15. Raven, J.A. (1975).
New Phytol. 74, 163-172

16. Rinaudo, G. (1983).
Ph.D. Thesis, Université de Paris-Sud, Centre d'Orsay

17. Döbereiner, R.J. and Day, J.M. (1976).
In "Proceedings of the 1st Int. Symp. on Nitrogen Fixation"Newton, E. and Nijman, C.J. (Eds.) 2, 518-538

18. Horemans, S., Van Onckelen, H.A., Rüdelsheim, P. and De Greef, J.A. (1984).
J. Exp. Bot., 35, 1832-1845

19. Hartmann, A., Fusseder, A. and Klingmüller, W. (1983).
In Klingmüller, W. (ed.) : Azospirillum II, Genetics Physiology Ecology. Experientia Suppl. 48, 78-88

AUXIN PRODUCTION BY AZOSPIRILLUM BRASILENSE UNDER DIFFERENT CULTURAL CONDITIONS

R. De Francesco*, G. Zanetti*, P. Barbieri** and E. Galli**

* Dipartimento di Fisiologia e Biochimica Generale - Sez. Chimica Biologica

** Dipartimento di Biologia - Sezione Genetica e Microbiologia

Università degli Studi di Milano, via Celoria 26 - 20133 Milano - Italy

Summary: Different *Azospirillum* strains grown in the presence of tryptophan were checked for auxin production by HPLC and TLC analyses. The highest levels of auxin were obtained both in N_2-fixing conditions and in limiting ammonia stationary cultures of strain Sp6. Two mutants of strain Sp6 were shown to be suitable for plant seedlings inoculation, to discriminate between the N_2-fixation effect and the hormonal effect on plant growth. Key words: auxin production, *Azospirillum* mutants.

Introduction

It is not yet known if the beneficial effects induced on crop plants by seeds inoculation with *Azospirillum* strains (1) exploiting the natural rhizocoenosis recently discovered (2), are due to the *Azospirillum* N_2-fixing capacity (3) or to the *Azospirillum* production of phytohormone-like substances (4) or possibly to both. Thus, the aim of the present study was to find *Azospirillum* mutants which were either nif^- and normal auxin-producers or nif^+ and lower auxin-producers, for inoculation of plant seedlings "in vitro".

Materials and methods

Chemicals. Indole-3-acetic acid (IAA) and indole-3-lactic acid (ILA) were from Sigma. Ethyl acetate, methanol and acetonitrile were Lichrosolv grade products from Merck; all other chemicals were analytical reagent grade.

Azospirillum III: Genetics, Physiology, Ecology
Ed. by W. Klingmüller

Bacteria and cultural conditions. The *Azospirillum* strains used in the present study are listed in Table 1.

Table 1. *Azospirillum brasilense* wild type and mutant strains

Strains	Source and main characteristics
A.brasilense Sp 6	wild type strain (5)
mutant SpF103	nif^- mutant of Sp6 (5)
mutant SpF57FT	fluorotryptophan-resistant mutant of Sp6 (6)
A.brasilense SpCd	ATCC No.29710, wild type strain
A.brasilense SpS631	reported not to produce auxin (7)

The bacteria were grown in a minimal mineral solution (pH 7) containing (in µg/liter): KH_2PO_4, 4.5; $K_2HPO_4 \cdot 3\ H_2O$, 1.8; $MgSO_4 \cdot 7\ H_2O$, 0.18; NaCl, 0.1; $NaMoO_4 \cdot 2\ H_2O$, 0.01; $CaCl_2$, 0.01; $FeCl_3 \cdot 6\ H_2O$ 0.0125; Na Succinate, 5. Different nitrogen sources were used. Medium A (limiting NH_4^+) contained 2.5 mM NH_4Cl; medium B (excess NH_4^+) contained 25 mM NH_4Cl; medium C (N-free) was without NH_4Cl and was used for growth under nitrogen-fixation conditions in Roux bottles. Cultures were incubated at 28°C under shaking when stated. Growth was measured as absorbance increase at 670 nm.

Isolation and determination of indole acetic acid. The procedure was according to Tien *et al.* (4) with slight modifications. The culture broth (50 ml), made cell-free by centrifugation and adjusted to pH 2.8, was extracted with ethyl acetate (50 ml x 2). The residue after evaporation under vacuum at 30°C, was dissolved in methanol and filtered through a 0.43 µm membrane filter. High performance liquid chromatography (HPLC) analysis of the methanol extracts was performed by isocratic reverse phase on a Waters Associates µBondapak C_{18} column (0.39 x 30 cm) equilibrated in water:acetonitrile:acetic acid (80:20:1) (vol/vol) at room temperature. Detection was at 280 nm and quantitation was made by area integration through the Waters Data System. Thin layer chromatography (TLC) was performed as described by Hartmann *et al.* (8) and developed with Ehrlich reagent (9).

Results and discussion

In Fig. 1 are reported for comparison chromatograms of an authentic sample of IAA and of an ethyl acetate extract of Azospirillum culture broth. Under the conditions used, HPLC analysis gave reliable results for indole acetic concentrations in the culture broth greater than 40 ng/ml. Thus, indole acetic acid was detectable only when tryptophan was added to the culture medium. The optimal concentration of L-tryptophan in the culture was 100 µg/ml.

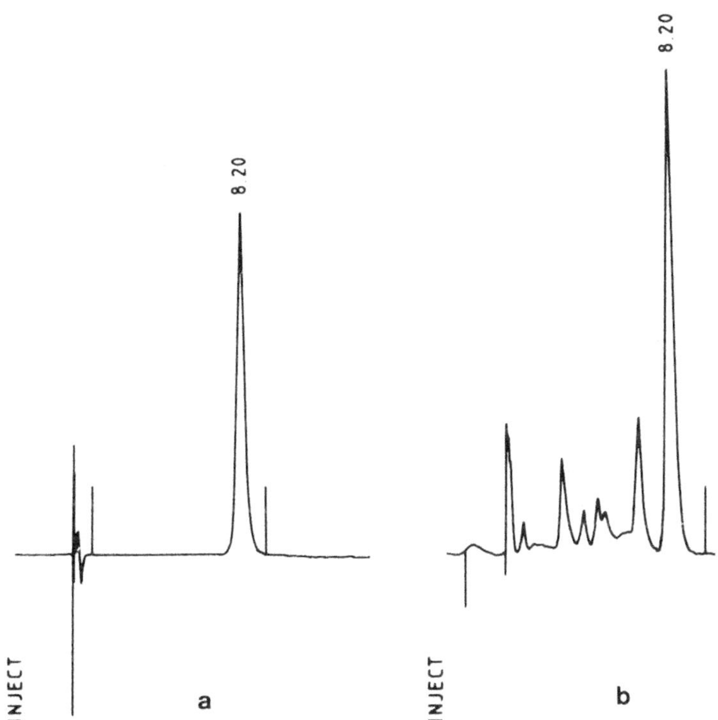

Figure 1. HPLC chromtograms of authentic IAA (a) and of the ethyl acetate extract of an A.brasilense culture broth (b).

In fig. 2 the production of indole acetic acid and of indole-3-lactic acid were followed, in a typical experiment, under limiting ammonium concentration in the presence of tryptophan. IAA production reached its maximum in the

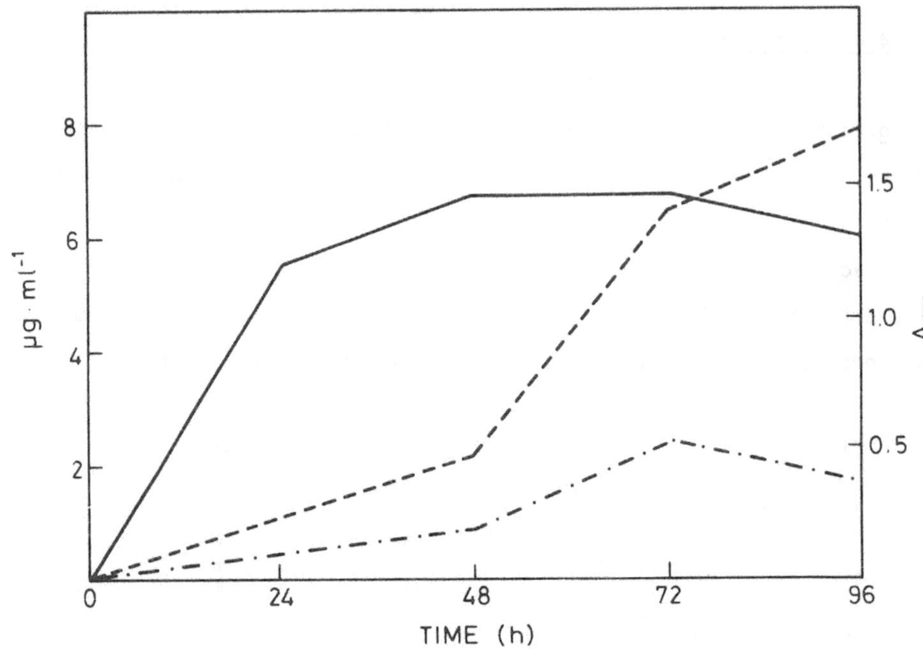

Figure 2. Time course of the production of IAA (---) and ILA (-·-) by A.brasilense Sp6. Growth (——) was in medium A (2.5mM NH_4^+) under shaking.

stationary phase. Indole-3-lactic acid was also measured, but due to the low amount formed and to difficulties in peak identification, its formation was not followed in further experiments.

The production of IAA by A.brasilense Sp6 under various growth conditions was followed. The time course of growth and IAA formation is reported in Fig.3. Clearly, the highest levels of auxin were obtained in N-free medium in the late log phase suggesting that tryptophan may be utilized as N source for growth. Instead, in all the other conditions the highest amounts of auxin were found in the stationary phase. Comparable levels of auxin were obtained both in N_2-fixing conditions and in limiting-ammonia stationary cultures. Such conditions probably are the nearest to the "in vivo" situation. Table 2A summarizes the data by comparing levels of auxin produced per A_{670} at the same culture age. By testing different A.brasilense mutants with nif$^-$ phenotype it was found that

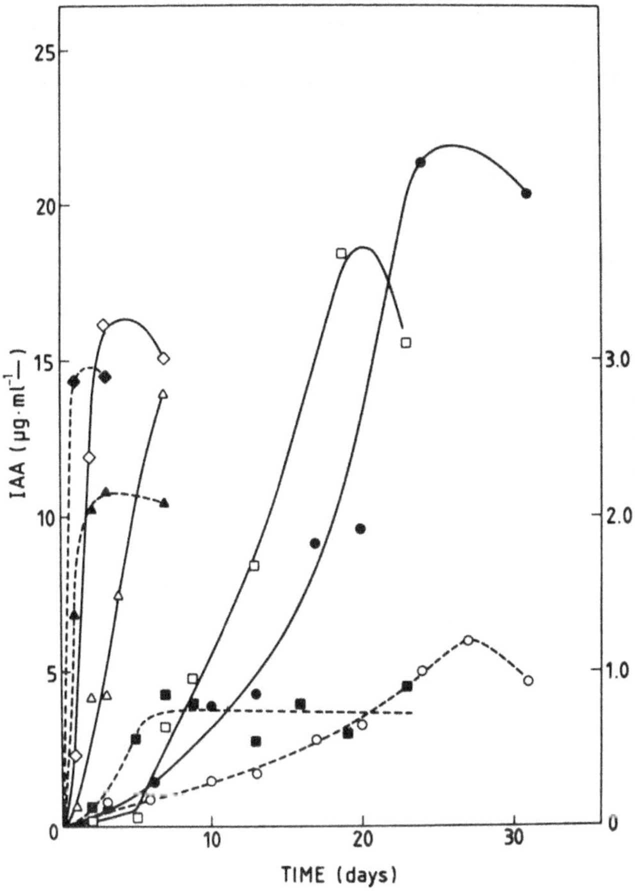

Figure 3. Time course of IAA production (——) and growth (---) by A.brasilense Sp6 under different cultural conditions. (◆,◇) medium B (25mM, NH_4^+), shaking; (▲,△) medium A (2.5mM, NH_4^+), shaking; (■,□) medium A (2.5mM, NH_4^+), stationary; (●,○) medium C (N_2-fixation), stationary.

strain SpF103 produced amounts of auxin higher than the wild type strain, as shown in the same table. It may be assumed that this mutant, which is also repressed in the glutamate synthase biosynthesis, could be altered in tryptophan metabolism. Alternatively, TLC analysis was utilized for a rapid screening of auxin production by different strains of Azospirillum. Results reported in Table 2B show that strains Sp6 and Cd have the same ability to produce large amounts of IAA in the presence of tryptophan; instead, strain SpS631 which was reported (7) not to produce auxin, formed in the same conditions a conspicuous amount of IAA. Strain SpF57FT, which is a fluorotryptophan resistant mutant, unexpectedly did not produce auxin in the absence of tryptophan, while in its presence the IAA produced was scarcely detectable. These data suggest that resistance to fluorotryptophan could be due to inefficient uptake of tryphophan. Thus, strains SpF103 and SpF57FT could be the mutants of choice for "in vitro" seedling inoculation in a system similar to that described by Bazzicalupo *et al.*, (10) but modified to allow much longer incubation times.

Table 2. Auxin production by different strains and mutants of *Azospirillum brasilense*

A) data were obtained by HPLC analysis of culture broth

B) data were obtained by TLC analysis of culture broth

A			B		
Strains	Growth conditions	IAA $\mu g/A_{670}/7$ days	Strains	Growth conditions	IAA
Sp6	N-free medium	9	Sp6	2.5 mM NH_4^+ ,shaking	++++
"	2.5 mM NH_4^+, stationary	4	SpCd	" " " "	+++
"	2.5 mM NH_4^+, shaking	7	SpS631	" " " "	++
"	25 mM NH_4^+, shaking	5	SpF57FT	" " " "	+
SpF103	25 mM NH_4^+, shaking	14			

Acknowledgements

This work was supported by the Consiglio Nazionale delle Ricerche, Italy.

References

1) Azospirillum II - (Klingmüller, W., ed) Exp. 48, Birkhäuser, Basel.

2) Döbereiner, J. and Day, J.M. (1976). In Proceedings of the 1st International Symposium on N_2 fixation (Newton, W.E. and Nyman, C.J. eds.) pp. 518-537, Washington State Univ. Press.

3) Döbereiner, J. (1983). In Azospirillum II (Klingmüller, W., ed). Exp 48, pp.9-23, Birkhäuser, Basel.

4) Tien, T.M., Gaskins, M.H. and Hubbel, D.H. (1979). Appl. Environ. Microbiol. 37, 1016-1024.

5) Bani, D., Barberio, C., Bazzicalupo, M., Favilli, F., Gallori, E. and Polsinelli, M. (1980). J. Gen. Microbiol. 119, 239-244.

6) Marocco, A., Bazzicalupo, and Perezin, M. (1983). In Azospirillum II (Klingmüller, W., ed.) Exp. 48, pp. 149-158. Birkhäuser, Basel.

7) Reynders, L. and Vlassak, K. (1979). Soil Biol. Biochem. 11, 547-548.

8) Hartmann, A., Singh, M., and Klingmüller, W. (1983). Can. J. Microbiol. 29, 916-923.

9) Bentley, J.A. (1962). Methods Biochem Anal. 9, 75-124.

10) Bazzicalupo, M., Cresta, E. and Favilli, F., (1985). Azospirillum III. In press.

COMPARISON OF NITROGENASE REGULATION IN A. BRASILENSE, A. LIPOFERUM, AND A. AMAZONENSE

A. Hartmann, H.-A. Fu, S.-D. Song and R. H. Burris
Department of Biochemistry, University of Wisconsin-Madison
College of Agricultural and Life Sciences
420 Henry Mall, Madison, Wisconsin 53706 U.S.A.

Summary: The influence of amino acids, ammonium chloride and oxygen on the nitrogenase activities of *Azospirillum* spp. was examined. In the presence of glutamate, aspartate, histidine or proline nitrogen fixation in *A. brasilense* strains was only slightly reduced, whereas in *A. lipoferum* and *A. amazonense* strains these amino acids almost completely abolished nitrogen fixation. Differences in the levels of ammonium assimilatory enzymes were found. After the addition of ammonium chloride, nitrogenase activity was completely inhibited in *A. brasilense* and *A. lipoferum*, but only partially inhibited in *A. amazonense*. In *A. brasilense* and *A. lipoferum* a covalent modification of the Fe protein of nitrogenase was involved in the "NH_4^+-switch off" as in *R. rubrum*. In *A. amazonense* no modified subunit of the Fe protein was found. Oxygen inhibited the nitrogenase activity of *A. brasilense* most effectively, whereas in *A. amazonense* the nitrogenase activity was remarkably oxygen tolerant. After the cells were exposed to inhibitory oxygen concentrations, the nitrogenase activity of all three species was recovered at a low pO_2. No modification of the Fe protein occurred during the "oxygen-switch off". However, under anaerobic conditions, the Fe protein of *A. brasilense* and *A. lipoferum* was modified.
Keywords: nitrogenase regulation, ammonium switch off, oxygen inhibition, *A. brasilense*, *A. lipoferum*, *A. amazonense*.

Introduction

An adequate supply of carbon and energy sources, absence of reduced nitrogen and an optimal oxygen concentration probably are limiting in the nitrogen fixation activity of *Azospirillum*. High concentrations of oxygen and fertilizer nitrogen may inhibit and/or repress nitrogen fixation. We were interested in differences in the regulation of nitrogen fixation in the

Azospirillum III: Genetics, Physiology, Ecology
Ed. by W. Klingmüller

Azospirillum spp., and we wanted to determine the regulatory mechanisms involved. In some photosynthetic, N_2-fixing bacteria, e.g., in R. rubrum, small amounts of ammonium chloride rapidly inhibit the nitrogenase activity - "NH_4^+-switch off" (1). A covalent modification of the Fe protein of nitrogenase occurs, which was recently characterized as ADP-ribosylation (2). An activating enzyme can reactivate the modified, inactive Fe protein by removal of the modifying group (3). The nitrogenase system of A. brasilense Sp 7 had been described as a similar three enzyme system - dinitrogenase, dinitrogenase reductase (Fe protein) and an activating enzyme (4,5). The general properties of the ammonium metabolism and the microaerobic behavior of A. brasilense Sp 7 (6,7) and A. lipoferum Sp Br 17 (8) have received some study. In this paper, we compare the regulation of nitrogen fixation by ammonium, amino acids and oxygen in A. brasilense, A. lipoferum and A. amazonense. In addition, the mechanisms of nitrogenase regulation are described and compared to other N_2-fixing organisms.

Material and Methods

The cultures of A. brasilense and A. lipoferum were from the American Type Culture Collection. A. amazonense Y 1 was kindly provided by Prof. N.R. Krieg. The bacteria were grown in the mineral salt medium described by Albrecht and Okon (9) for A. brasilense and A. lipoferum and by Magalhaes et al. (10) for A. amazonense. For the microaerobic growth with N_2, semisolid medium or liquid cultures with dissolved O_2 in equilibrium with 0.003 atm O_2 (= 0.3 kPa O_2) controlled by an oxystat were used. The pH was kept at 6.8 with a pH-stat in A. brasilense and A. lipoferum and at 6.0 in A. amazonense cultures. Malate (0.5%) was the carbon source in all cultures.

Nitrogenase activity was measured with the acetylene reduction assay. The semisolid cultures were stoppered and incubated at 30° under stagnant conditions for two hours. The acetylene reduction of liquid cultures was determined in a well-mixed chamber at 30° with an O_2-probe to measure dissolved oxygen concentrations simultaneously (11). In vitro nitrogenase assays were performed with the dithionite-dependent acetylene-reduction assay.

To prepare cell free extracts, 10 ml of N_2-fixing culture was filtered rapidly and frozen in liquid N_2; "quick extracts" of the cells on the filter were prepared according to Kanemoto and Ludden (1). After SDS-polyacrylamide gel electrophoresis with a low concentration of crosslinker (1), the proteins were transferred to a nitrocellulose filter by electrotransfer. Immunoblotting was performed with antiserum against the Fe protein of R. rubrum, kindly provided by Dr. P.W. Ludden. Goat anti-rabbit IgG horse-radish peroxidase conjugate (BioRad) was used as the second antibody to visualize the antigen in an enzyme-linked colorimetric reaction.

Results and Discussion

Influence of amino acids on the nitrogen fixation activity

Aspartate, glutamate, histidine and proline affected the nitrogen fixation of Azospirillum spp. differently. A. brasilense strains had high nitrogenase activity in the presence of these amino acids, whereas little or no nitrogenase activity was found in A. lipoferum and A. amazonense strains (Fig. 1 and Table 1). A. lipoferum and A. amazonense grew equally well on glutamate or ammonium, whereas A. brasilense grew much slower on glutamate than on ammonium. In addition, A. lipoferum and A. amazonense could readily use these amino acids as sole sources of carbon and nitrogen. Under these conditions ammonium was excreted (Table 1). Comparing the ammonium assimilatory enzymes of glutamate-grown azospirilla indicated that A. lipoferum and A. amazonense had higher glutamate dehydrogenase and glutamate uptake activities than A. brasilense. In contrast to A. lipoferum and A. amazonense, glutamate grown A. brasilense had active glutamine synthetase and NH_4^+-transport activities, indicating that the cells were N-limited. When glutamate (1 mM) was added to nitrogen-fixing cells, the nitrogenase activity of A. lipoferum and A. amazonense was partially inhibited after a lag phase of about 20 minutes; A. brasilense was not significantly affected under these conditions.

The different inhibitory effects of certain amino acids on N_2 fixation of Azospirillum spp. probably is due to a different efficiency in amino acid utilization. The efficient utilization of amino acids indicates a

Table 1. Amino acids and nitrogen fixation of *Azospirillum* spp.

		Azospirillum brasilense				*Azospirillum lipoferum*				*Azospirillum amazonense*
		Sp7	Sp6	SpCd	Sp35	SpRG59a	SpRG20a	SpBR17	SpRG6xx	Y1
A) Growth on malate:										
	ABS	1.7	1.7	2.1	2.1	1.2	1.7	1.9	1.9	0.4
	ARA	57	64	31	33	42	40	35	25	36
	NH_4^+	0	0	0	0	0	0	0	0	0
B) Growth on malate with amino acids (10 mM) present:										
glutamate:	ABS	0.93	0.80	2.8	2.2	3.4	3.2	3.8	3.2	0.97
	ARA	46	28	17	2.6	0	0	0	0.08	1.1
	NH_4^+	0	0	0	0	0.70	0.88	0.05	0.92	0.03
histidine:	ABS	1.5	1.2	1.7	1.8	3.2	3.5	3.6	3.5	0.90
	ARA	55	55	36	33	0	0	0	0	0
proline:	ABS	3.9	2.8	4.7	4.0	3.5	4.2	4.6	4.2	2.6
	ARA	1.2	27.0	4.4	1.3	0	0	0	0.7	0
	NH_4^+	0	0	0	0	0.03	0.05	0.05	0	0.10
C) Growth with amino acids (20 mM) as sole N- and C-source:+)										
glutamate:	ABS	0.39	0.60	0.27	1.62	3.0	3.0	3.0	2.8	0.64
	NH_4^+	0.05	0.06	0.04	0.28	11.8	11.4	11.8	12.2	0.26
histidine:	ABS	0.02	0.01	0.02	0.01	0.42	1.38	0.40	0.16	0.07
proline:	ABS	0.25	0.15	0.17	0.26	2.5	3.4	2.9	2.8	2.2
	NH_4^+	0	0	0	0	0.66	1.2	1.2	0.46	1.05

Growth was performed at 30°, stagnant conditions in semisolid medium for 48 hours, before the acetylene reduction activity (nmol ethylene $hour^{-1} ml^{-1}$ [ABS=1], "ARA"), growth (absorbance at 580 nm "ABS") and extracellular ammonium (mM, "NH_4^+") was determined.

+) No nitrogenase activity was found under these conditions.

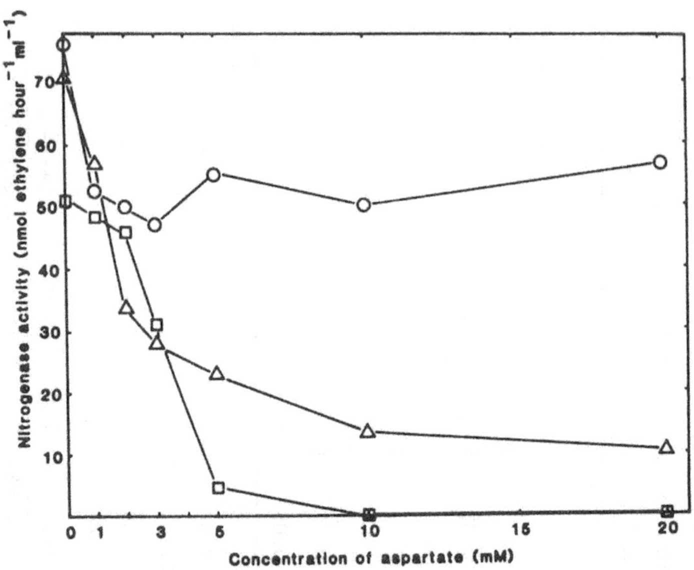

Figure 1. Influence of aspartate on nitrogen fixation of A. brasilense Sp 7 (o), A. lipoferum Sp RG 20a (□) and A. amazonense Y 1 (Δ) in semisolid medium. After incubation at 30° for 24 hours (A. amazonense 48 hours) with different initial concentrations of aspartate, the acetylene reduction activity was measured.

saprophytic life style for some Azospirillum strains on plant root exudates, and this could be a drawback for their effectiveness as N_2-fixing biofertilizers.

Regulation of nitrogenase activity by ammonium chloride

Ammonium chloride inhibited the nitrogenase activity of all Azospirillum spp. reversibly at 50-100 μM concentration (NH_4^+-switch off/on). When the "switch off" with 1 mM NH_4Cl was performed under oxygen-controlled conditions, the inhibition of nitrogenase was complete after a transition period in A. brasilense and A. lipoferum, but it was not complete in A. amazonense (Fig. 2). Even 10 mM ammonium chloride did not completely inhibit its nitrogenase activity. In A. brasilense and A. lipoferum the in vitro-

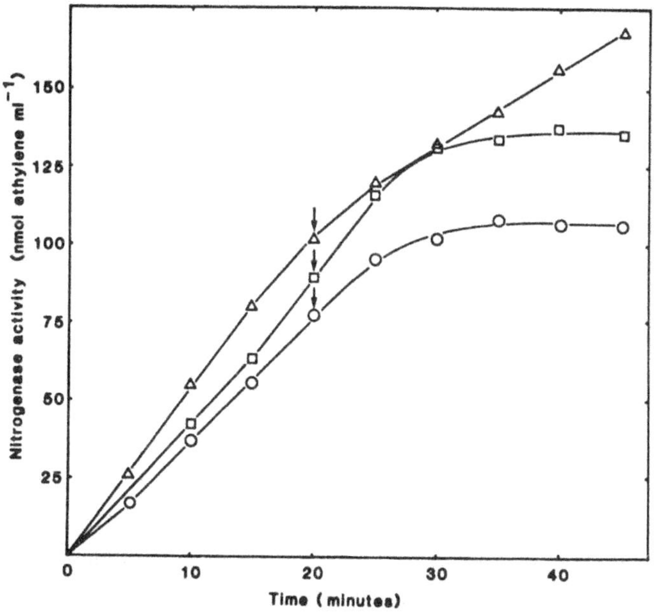

Figure 2. "NH_4^+-switch off" of nitrogenase activity in A. brasilense Sp 7 (o), A. lipoferum Sp RG 20a (□) and A. amazonense Y 1 (Δ). The acetylene reduction activity was measured in an oxygen chamber at a dissolved oxygen concentration of 0.001 atm O_2 (0.1 kPa). At 20 minutes (arrows) 1 mM NH_4Cl was added.

nitrogenase activity was reduced after the addition of ammonium chloride to the cells; in A. amazonense the in vitro nitrogenase activity was not affected by ammonium chloride.

To elucidate the molecular mechanism of the "NH_4^+-switch off", we prepared "quick extracts" of N_2-fixing cultures before and after NH_4^+-treatment and examined the status of the Fe protein with the immunoblotting technique. In A. brasilense and A. lipoferum an additional Fe protein subunit appeared during the "NH_4^+-switch off", and it migrated like the modified subunit of the inactive Fe protein of R. rubrum (Fig. 3). An incorporation of ^{32}P into this subunit was observed in a ^{32}P-labelled culture of A. brasilense Sp 7. In A. amazonense, a closely migrating double band

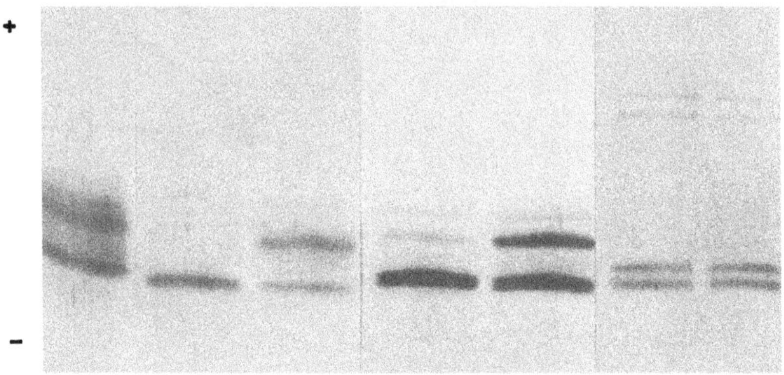

Figure 3. Immunoblots from "quick extracts" prepared from cultures of A. brasilense Sp 7 (A), A. lipoferum Sp RG 20a (B) and A. amazonense Y 1 (C) before (lanes 1) and after (lanes 2) the addition of 1 mM NH_4Cl. Rr2i: inactive Fe protein of R. rubrum.

of Fe protein was present before and after "NH_4^+-switch off" (Fig. 3), but no additional modified Fe protein subunit appeared. The same protein pattern of Fe protein subunits was found in highly active, purified nitrogenase of A. amazonense (Song, Hartmann, Burris, submitted).

The regulation of nitrogenase by ammonium chloride in A. brasilense and A. lipoferum is similar to the regulation in R. rubrum. Covalent modification of the Fe protein produces reversible inactivation of nitrogenase. In A. amazonense, the weak inhibition of nitrogenase by ammonium chloride and the lack of modification of the Fe protein resembles the control situation in Azotobacter vinelandii. The rather weak inhibition of nitrogenase by ammonium chloride in A. amazonense is an advantage of this bacterium when functioning as a provider of N biofertilizer.

Regulation of nitrogenase activity by oxygen

The influence of dissolved oxygen on the nitrogenase activity of Azospirillum spp. was measured in a specially designed reaction vessel (11), that allowed the simultaneous measurement of dissolved oxygen and of acetylene reduction activity. A rapidly stirring magnetic bar provided quick equilibration between the gas and liquid phase. Figure 4 shows a comparison of the oxygen profiles of the nitrogenase activities for three species of azospirilla. The optimal range for nitrogen fixation by Azospirillum spp. was 0.001-0.002 atm O_2 (0.1-0.2 kPa). The nitrogenase activity of A. brasilense Sp 7 was inhibited more effectively by oxygen than it was in A. amazonense Y 1. The inhibition by oxygen was partially

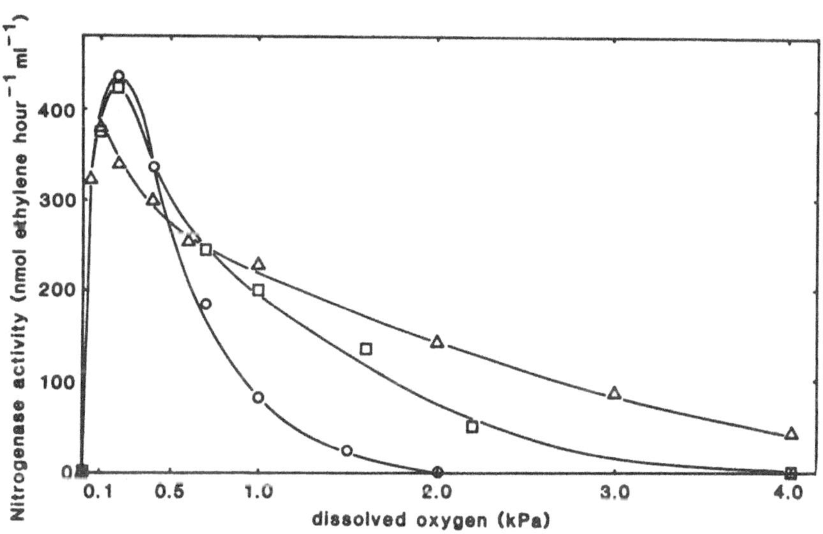

Figure 4. Influence of dissolved oxygen on the nitrogenase activity of A. brasilense Sp 7 (o), A. lipoferum Sp RG 20a (□) and A. amazonense Y 1 (Δ). Culture samples (4 ml) of microaerobically grown cultures (0.3 kPa dissolved oxygen) were transferred to an oxygen chamber, and acetylene reduction was measured at different dissolved oxygen concentrations.

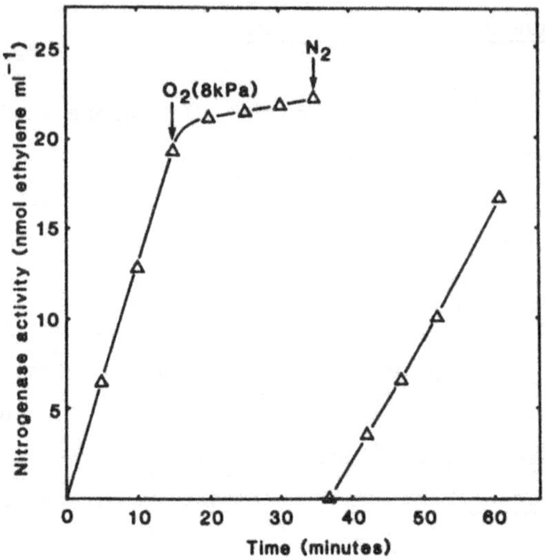

Figure 5. "Oxygen-switch off" of nitrogenase activity in A. amazonense Y 1. Acetylene reduction was measured in the oxygen chamber at 0.2 kPa dissolved O_2 before the shift (arrow) to 8 kPa oxygen. Ten minutes afterwards, the chamber was flushed with nitrogen, the dissolved oxygen was reset to 0.2 kPa, and acetylene (0.1 atm) was injected to continue the assay (return to baseline).

reversible in all Azospirillum spp. For example, the "oxygen-switch off" of nitrogenase activity in A. amazonense Y 1 by 0.08 atm O_2 and its reversal is shown in Fig. 5.

To characterize the mechanism involved, the Fe protein was examined in "quick extracts" of rapidly filtered cells with the immunoblotting method. No modification of the Fe protein was found in cells with oxygen-inhibited nitrogen fixation. The oxidation of reducing equivalents for nitrogenase may have caused the inhibition. Under anaerobic conditions, which also inhibit the nitrogenase activity of Azospirillum spp. reversibly, the Fe proteins of A. brasilense and A. lipoferum were modified.

Since the modification of the Fe protein in A. brasilense and A. lipoferum is involved in the "anaerobic inhibition" and the "NH_4^+-switch off" of nitrogenase, the modifying enzyme of the Fe protein may be triggered by two different signals. The covalent modification system clearly has the function to prevent waste of energy in the nitrogenase reaction.

There are clear differences in the oxygen tolerance among the Azospirillum spp; A. amazonense Y 1 again appears as a favorable organism because of its relatively high tolerance to oxygen. The oxygen protective mechanisms involved deserve further investigations.

Acknowledgements

This work was supported by the College of Agricultural and Life Sciences, University of Wisconsin-Madison and by grants from the National Science Foundation and by the U.S. Department of Agriculture. A. H. also was supported by the Deutsche Forschungsgemeinschaft.

We greatly appreciate the generous support and helpful discussions with Drs. P.W. Ludden, L.L. Saari and R.H. Kanemoto.

References

1. Kanemoto, R.H. and Ludden, P.W. (1984). J. Bacteriol. 158, 713-720
2. Pope, M.R., Murrell, S.A. and Ludden, P.W. (1985). Proc. Natl. Acad. Sci. USA 82, 3173-3177
3. Saari, L.L., Triplett, E.W. and Ludden, P.W. (1984). J. Biol. Chem. 259, 15502-15508
4. Ludden, P.W., Okon, Y. and Burris, R.H. (1978). Biochem. J. 173, 1001-1003
5. Pedrosa, F.O. and Yates, M.G. (1983). In Klingmüller, W. (ed.): Azospirillum, Genetics Physiology Ecology. EXS 48, pp 66-77, Birkhäuser, Basel

6. Okon, Y., Albrecht, S.L. and Burris, R.H. (1976). J. Bacteriol. 128, 592-597
7. Okon, Y., Houchins, J.P., Albrecht, S.L. and Burris, R.H. (1977). J. Gen. Microbiol. 98, 87-93
8. Volpon, A.G.T., De-Polli, H. and Döbereiner, J. (1981). Arch. Microbiol. 128, 371-375
9. Albrecht, S.L. and Okon, Y. (1980). In San Pietro, A., (ed.): Methods in Enzymology, 69, pp 740-749. Academic Press, New York
10. Magalhaes, F.M., Baldani, J.I., Souto, S.M., Kuykendall, J.R. and Döbereiner, J. (1983). An. Acad. Brasil. Cienc. 55, 417-430
11. Hochman, A. and Burris, R.H. (1981). J. Bacteriol. 147, 492-499

SOME PHYSIOLOGICAL AND BIOCHEMICAL PROPERTIES OF DENITRIFICATION BY AZOSPIRILLUM BRASILENSE

G. Danneberg, W. Zimmer and H. Bothe
Botanisches Institut, Universität Köln
Gyrhofstr. 15, D-5 Köln 41

Summary: Previous experiments from this laboratory showed that *Azospirillum brasilense* Sp 7 can grow anaerobically with either nitrate, nitrite or nitrous oxide as respiratory electron acceptors. Molar growth yield measurements now indicate that *Azospirillum* obtains the same amount of energy with NO_2^- and O_2 as respiratory electron acceptors but only 1/3 of that with NO_3^-. Dissimilatory nitrite reductase catalyzing the reduction of NO_2^- to N_2O is shown to be a cytochrome c, d containing enzyme. A soluble dissimilatory nitrite reductase preparation also catalyzes the formation of nitric oxide which is probably an artifact of the cell-free system. The expression of dissimilatory N_2O-reductase in *Azospirillum* is dependent on Cu in the medium.
Key words: dissimilatory nitrate reduction, nitrite reduction, denitrification, nitrous oxide, nitric oxide, nitrite.

Introduction

The role of *Azospirillum* in enhancing crop productivity has extensively been discussed in the preceeding two workshops on *Azospirillum*. The bacterium may provide excess of fixed nitrogen formed by N_2-fixation to the roots of cereals. Alternatively it could stimulate the growth of the roots and their mineral uptake by excreting phytohormones when living on the surface or inside of the roots (1,2). Generally, *Azospirillum* is only discussed as being beneficial to crop plants. However, the bacterium may cause just the opposite under unfavorable conditions. Similar as *Rhizobium* (3), *Azospirillum* can utilize either nitrate,nitrite or nitrous oxide as respiratory electron acceptor (4,5, 6, 7). The final products of these dissimilatory reductions can be either

Azospirillum III: Genetics, Physiology, Ecology
Ed. by W. Klingmüller

nitrite or gaseous nitrous oxide or dinitrogen (6,7). Model experiments with wheat-Azospirillum associations showed that Azospirillum, indeed, performs either N_2-fixation (C_2H_2-reduction) or denitrification (N_2O-formation), depending on the amount of nitrate and/or oxygen available(8). Both activities were strictly dependent on the growth temperature of the association and on the Azospirillum strain used (8). The Azospirillum strains available are not yet well characterized with respect to denitrification. Some strains appear to reduce nitrate only to nitrite, whereas others can convert nitrate via nitrite to nitrous oxide and molecular nitrogen (9). The physiological conditions for maximal gas production are partly understood (6,7). Likewise, the biochemical properties of the enzymes involved in denitrification have not yet been elucidated for Azospirillum.

Our group is currently trying to characterize the physiological and biochemical properties of denitrification by Azospirillum. Some results on nitrate, nitrite and nitrous oxide dissimilation will be presented here.

Materials and Methods

Azospirillum brasilense Sp 7 (ATCC 29145) was used in the present investigation. For growth yield measurements, Azospirillum was grown in a chemostatic culture(6) with L- malate limiting. Details are described elsewhere (W. Zimmer, Diplomarbeit, Köln, 1984). For preparing a cell-free preparation performing dissimilatory nitrite reduction, bacteria were used which had been grown anaerobically in continuous culture with nitrite as the respiratory electron acceptor (6) and with DL-malate as the carbon source. The cells were harvested by centrifugation (10 min, 16 000 g), suspended in 10 mM K-phosphate buffer pH 7.5 and broken twice in a French Press at 140 000 kPa. The extract was centrifuged (30 min, 12 000 g) and the supernatant was spun down once more (2h, 100 000 g). The pellet was then suspended in 10 mM K-phosphate buffer pH 7.5 and assayed for activity. All preparations were performed in air. The assays were performed in 7.0 ml Fernbach flasks containing in a final volume of 3 ml: extract with 0.54 mg protein, Na-nitrite 20 µmol, phenazine methosulfate 0.8 µmol, K-phosphate buffer pH 7.5 500 µmol. The assays were started by adding 50 µmol Na-ascorbate. The reaction was performed for 45 min at 30°C under argon in a shaking water bath. The amount of N_2O formed was determined by gas chromatography. NO formation was followed amperometrically in a separate assay by using a conventional Clark-type-electrode (10). Nitrite was determined colourimetrically with the naphthylamine/sulphanilic acid reagent.

For the growth of Azospirillum under anaerobic conditions with N_2O as the respiratory electron acceptor, a batch culture grown with NH_4Cl (5.0 mM) as the nitrogen source was used as the starting culture. Growth of the cells under different conditions was followed in 100 ml flasks containing Azospirillum NH_4^+-medium, EDTA and the different salts tested as indicated in Table 3 and Fig.2. The flasks were inoculated with 0.1 ml of the starting culture (O.D. 560 nm = 0.06) and evacuated and gassed with argon. The assays were started by adding 20 % (v/v) N_2O, and the cells were grown for 16 h at 30°C in a shaking water bath. Cell growth was monitored by measuring the optical density at 560 nm.

Results and Discussion

Previous results had shown that Azospirillum grows in continuous culture with either O_2, nitrate or nitrite as respiratory electron acceptor (6). Table 1 compares the maximal molar growth yields (Y_S^{max}, determined for infinite dilution rates in continuous cultures) which we now determined for the respiratory electron acceptors O_2,NO_3^- and NO_2^-. The maximal growth yield is 48 g cells/mol malate consumed in the case of O_2 and NO_2^-, but only 15.3 with NO_3^-. Such findings are not surprising, because the free energy change ($\Delta G_0'$) is only 162 KJ/mol for the reduction of nitrate to nitrite by NADH, but is 220 or 229 KJ/mol for the reduction of O_2 or NO_2^-, respectively, by NADH (11, see Table 1). The data indicate that the same amount of energy (ATP) is obtained when O_2 or NO_2^- is used as respiratory electron acceptor, but only 1/3 of that with nitrate. The same conclusion follows from other independent measurements which will be published elsewhere (Zimmer and Bothe, in preparation).

Table 1. Molar growth yields (Y_S^{max}) of a malate grown, chemostatic culture of Azospirillum brasilense Sp 7 using different respiratory electron acceptors

Respiratory electron acceptor:	O_2	NO_2^-	NO_3^-
Y_S^{max} (g/mol):	48	48	15.3
ratio Y_S^{max}: Y_S^{max} of the O_2-culture:	---	1:1	0.32:1
$G^{o\prime}$ (KJ/mol) *	220	229	162

* The $\Delta G^{o\prime}$- values are taken from Thauer et al. (11) and refer to the reactions:

a) for O_2: $1/2\ O_2 + NADH + H^+ \longrightarrow H_2O + NAD^+$
b) for NO_2^-: $NO_2^- + NADH + 2H^+ \longrightarrow 3/2\ H_2O + 1/2\ N_2O + NAD^+$
c) for NO_3^-: $NO_3^- + NADH + H^+ \longrightarrow H_2O + NO_2^- + NAD^+$

Two different types of dissimilatory nitrite reductases have been described for denitrifying bacteria (12,13). The more commonly distributed enzyme (occurring in Paracoccus denitrificans, Pseudomonas strains and Thiobacillus denitrificans, e. g.) contains cytochrome c,d in the prosthetic group and is blocked by NH_2OH. A non-heme dissimilatory nitrite reductase containing Cu is produced by Alcaligenes faecalis or Rhodopseudomonas sphaeroides f. denitrificans. The latter enzyme catalyzes the reduction of NH_2OH. Previous experiments with intact cells showed that NH_2OH blocks dissimilatory nitrite reduction by Azospirillum suggesting that this bacterium possesses a cytochrome c, d type enzyme (6). This can directly by shown in a preparation from *Azospirillum brasilense* Sp 7. The preparation was obtained by centrifuging the crude extract at 100 000 g for 2 h, and the supernatant was then fractionated by $(NH_4)_2SO_4$ precipitation. The fraction precipitating between 55 and 90 % $(NH_4)_2SO_4$ could be reduced by phenazine methosulfate/NADH and reoxidized by $NaNO_2$.The difference spectra (reduced by phenazine methosulfate/NADH minus oxidized as obtained after the isolation =straight line) or reduced by phenazine methosulfate/NADH and reoxidized by $NaNO_2$ minus oxidized as obtained after the isolation = dashed line) revealed the presence of six different bands (Fig 1). The bands at 415, 519 and 548 nm might be associated with cytochrome c and those at 460 , 606 and 650 nm are indicative for the occurrence of cytochrome d (cytochrome c_2) (see 14, 15, 16). Thus *Azospirillum* possesses a cytochrome c,d dissimilatory nitrite reductase. The bacterium may also contain a Cu-enzyme which has not yet been investigated. However, this would be an unusual feature because no bacterium has been described up-till-now which contains both types of dissimilatory nitrite reductases.

Dissimilatory nitrite reductase should be membrane-bound, because it is functioning in respiratory energy generation (6). Table 2 compares the specific activities of dissimilatory nitrite reductase in crude extracts and in the particulate and soluble fraction obtained after centrifugation at high speed. The crude extract significantly reduced nitrite. However, the stoichiometry between N_2O-formation and NO_2^--disappearance was far from the expected relation (1:2). Thus other products besides N_2O were likely produced from nitrite. The crude extract did not reduce N_2O further to N_2. This could be shown by adding C_2H_2 which specifically blocks N_2O-reductase in denitrifying bacteria (see 7). The addition of C_2H_2 did not enhance the rate of N_2O-formation (see Table 3). Crude extracts, however, converted nitrite to nitric oxide to a large extent. When the extract was centrifuged, most of the NO_2^--reducing activity was recovered in the supernatant (Table 2). The supernatant containing the soluble dissimilatory nitrite

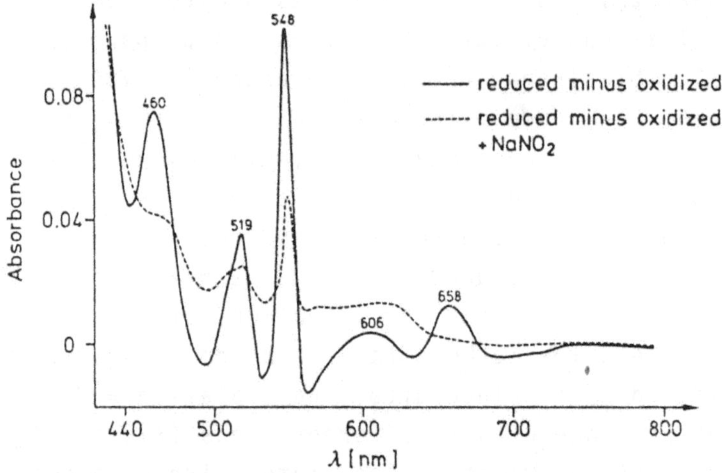

Fig. 1. Difference spectra of the solubilized dissimilatory nitrite reductase enriched by ultracentrifugation and $(NH_4)_2SO_4$-precipitation

straight line: difference spectrum: enzyme reduced by NADH and phenazine methosulfate minus oxidized enzyme (in the reference cuvette)

dashed line: difference spectrum: enzyme reduced by NADH and phenazine methosulfate and reoxidized by $NaNO_2$ minus oxidized enzyme (in the reference cuvette)

For this assay, the crude extract obtained by French Press treatment was centrifuged (2 h, 100 000 g), and the supernatant was fractionated by $(NH_4)_2SO_4$. The fraction precipitating between 55 and 90 % $(NH_4)_2SO_4$ was dissolved in 10 mM K-phosphate buffer pH 7.5 and dialyzed overnight against the same buffer. The spectra from this preparation were recorded in a Perkin-Elmer 555 spectrophotometer. The cuvette contained for the reduction of the nitrite reductase: NADH 0.5 mM, phenazine methosulfate 0.025 mM, K-phosphate buffer pH 7.5 50 mM and protein 0.7 mg. The gas phase was argon. $NaNO_2$ (final concentration 12.5 mM) was added to the cuvette for the reoxidation of the sample.

Table 2. The distribution of dissimilatory nitrite reductase activity after ultracentrifugation

I. Specific activity:
(in nmol/ min x mg protein)

	NO_2^- consumption	N_2O formation	NO formation	ratio $NO_2^-:N_2O$	ratio $NO_2^-:NO$
a) crude extract	354	42.2	118	8.4:1	3.0:1
b) supernatant after ultracentrifugation	415	22.2	97	18.7:1	4.3:1
c) pellet after ultracentrifugation	131	38.9	28	3.4:1	4.7:1

II. Total activity:
(in nmol/ min)

	protein content	NO_2^--consumption		N_2O-formation		NO-formation	
	mg	total	%	total	%	total	%
a) crude extract	69.8	24556	100	2947	100	7403	100
b) supernatant after ultracentrifugation	54.6	22667	92	1210	41	5288	71
c) pellet after ultracentrifugation	13.6	1778	7	528	18	382	5

The crude extract (15 ml, obtained by French Press treatment at 140 000 KPa, 2x and centrifugation at 12 000 g, 30 min) was centrifuged (2 h, 100 000 g). The supernatant (11 ml) was directly assayed and the pellet was suspended in 10 mM K-phosphate buffer pH 7.5 (4ml). The preparations were assayed for NO_2^--reduction using phenazine methosulfate and Na-ascorbate as electron donors

Table 3. Dissimilatory formation of nitrous oxide by a membrane fraction from Azospirillum brasilense Sp 7

Assay condition	nmol N_2O formed/ min x mg protein
1. complete	68.4
2. complete + 2 % C_2H_2	69.6
3. complete, but gas phase air	17.1
4. - $NaNO_2$	3.4
5. - phenazine methosulfate	1.7
6. - Na-ascorbate	0
7. - extract	0
8. complete, but extract boiled	0

The membrane fraction was assayed for the phenazine methosulfate and Na - ascorbate dependent formation of nitrous oxide from nitrite. The pellet obtained after ultracentrifugation (2 h, 100 000 g) was assayed for activity. The exact assay conditions are described under Materials and Methods

reductase fraction produced much NO, but rather little N_2O. The sum of the NO and N_2O productions could not account for the nitrite disappearance, indicating that another, still unidentified product must have been formed during nitrite reduction. By contrast, the overall activity in the pellet was low. However, the stoichiometry between NO_2^--disappearance and N_2O-formation approached to 2:1, and much smaller amounts of NO were formed in the pellet than in the supernatant(Table 2).

Nitric oxide formation from nitrite is believed to be an artifact of cell-free preparations in Pseudomonas denitrificans (17), and this is presumably also the case in Azospirillum. Dissimilatory nitrite reductase appears to fall off the membranes easily, and the enzyme catalyzes the formation of free nitric oxide only in the solubilized form.

The dissimilatory nitrite reduction in the pellet was characterized further (Table 3). This preparation catalyzed the formation of N_2O which was strictly dependent on nitrite and on the reductants sodium ascorbate and phenazine methosulfate. The reaction required anaerobic conditions and proceeded linearly for at least 45 min and 1 mg protein/vessel. We are currently trying to characterize this preparation in more detail.

Dissimilatory nitrous oxide reductase was reported to be a copper enzyme (18,19) or not to contain this metal (20). Up-till now, this enzyme has only been investigated biochemically from the marine _Pseudomonas perfectomarinus_. _Azospirillum_ grows anaerobical-

Table 4. Growth of Azospirillum brasilense Sp 7 under anaerobic conditions with N_2O as the respiratory electron acceptor: Dependence on the metal-ion in the medium

	EDTA (300 μM)	salt (250 μM)	after 16 h ($O.D._{560nm}$)	association constant of the metal-EDTA complex ($\log K_S$)
1.	---	---	0.266	---
2.	+	---	0.018	---
3.	+	$CuSO_4$	0.173	18
4.	+	$CuCl_2$	0.165	18
5.	+	$CaSO_4$	0.027	9
6.	+	$CaCl_2$	0.029	9
7.	+	$NiCl_2$	0.069	18
8.	+	$CoSO_4$	0.074	16
9.	+	$ZnSO_4$	0.114	16
10.	+	$MgSO_4$	0.019	9
11.	+	$MnCl_2$	0.067	13.5

The growth and assay conditions are described under Materials and Methods

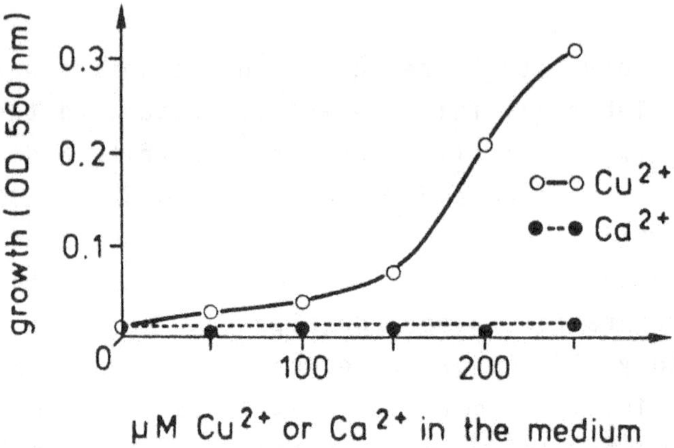

Fig. 2. Dependence of the growth of EDTA-treated *Azospirillum* on the presence of $CuSO_4$ or $CaSO_4$ in the medium

The experimental details are described under Materials and Methods (last section)

ly with N_2O as the sole respiratory electron acceptor. The bacterium must synthesize nitrogenase to meet the requirements for nitrogen unless the medium is supplemented with NH_4Cl(7). N_2O-dependent growth was further investigated in the present study. When the medium was supplemented with 300 μM EDTA, growth stopped (Table 4). EDTA did not unspecifically affect growth. This follows from the findings that Cu^{++} and to some extent Zn^{++}, but virtually not Mg^{++},Ca^{++}, Ni^{++} or Mn^{++} restored growth (measured by the increase in turbidity which had resulted 16 h after the inoculation) (Table 4). The stability constant for the Cu-EDTA-complex is similar or higher than those for the other metal-EDTA-complexes (21, see Table 3). This means that Cu is not less effective in binding EDTA and thus in removing the chelator out of the medium than other metals. The effects of the different salts are, therefore, not due to differences in the concentrations of free EDTA in the assays, but are caused by the metal-cations itself. It is, therefore, safe to conclude that Cu is involved in the expression of dissimilatory nitrous oxide reduction by *Azospirillum*. Fig. 2 indicates that 250 μM $CuCl_2$ was saturating for maximal growth under the conditions employed.

Concluding remarks

Our attempts to characterize denitrification by *Azospirillum* biochemically are still at an initial state. However, the present investigation showed that dissimilatory nitrite reductase is a cytochrome c,d containing enzyme. Further investigations have to elucidate the natural donor(s) for dissimilatory nitrate, nitrite and nitrous oxide reductases and the biochemical properties of the enzymes involved. Experiments are currently performed to measure directly the amount of ATP formed and the coupling to H^+-uptake in vesicle preparations. From the molecular growth yield measurements reported here it is likely that NO_2^- and O_2-respiration provide 3 ATP/2e whereas NO_3^- -respiration yields only 1 ATP/2 e transferred. The present investigation also showed that Cu is involved in the expression of N_2O-reduction activity in *Azospirillum*. This means that Cu is a constituent of the dissimilatory N_2O-reductase itself or is required for the biosynthesis of a factor involved in dissimilatory N_2O-reduction (e. g. an electron donor). It is planned to purify nitrous oxide reductase under anaerobic conditions similar as performed with the enzyme from *Pseudomonas perfectomarinus* (19). This should allow to show the presence of Cu in the prosthetic group of the enzyme.

Acknowledgement. This work was kindly supported by grants from the Bundesministerium für Forschung und Technologie (PTB Jülich).

References

1. Okon,Y.(1984). In Veeger, C. and Newton, W.E. (eds.): Advances in Nitrogen Fixation, pp. 303-309, Nijhoff-Junk-Pudoc (Publishers), The Hague,Boston, Lancaster.

2. Döbereiner, J.(1983). In Klingmüller,W.(ed.) Azospirillum II, Genetics, Physiology,Ecology, EXS 48,pp. 9-23,Birkhäuser,Basel

3. Van Berkum,P. and Kayser,H.H. (1985). Appl. Environ.Microbiol. 49, 772-777.

4. Nelson,L.M. and Knowles,R. (1978). Can. J. Microbiol. 24,1395-1403.

5. Tibelius, K.H. and Knowles, R. (1984). J. Bact. 157, 84 -88.

6. Zimmer, W.,Penteado Stephan,M. and Bothe, H. (1984)·Arch. Microbiol. 138, 206-211.

7. Penteado Stephan,M., Zimmer, W. and Bothe, H. (1984).Arch. Microbiol. 138, 212-216.

8. Neuer, G., Kronenberg, A. and Bothe, H. (1985).Arch. Microbiol. 141, 364-370.

9. Tarrand J.J., Krieg, N. R., Döbereiner, J. (1978).Can. J. Microbiol. 24, 967-980.

10. Zimmer W., Danneberg, G. and Bothe, H. accepted by Current Microbiol.

11. Thauer, R.K., Jungermann, K. and Decker,K. (1977).Bacteriol. Reviews 41, 100-180.

12. Bryan, B.A. (1981). In Delwiche, C. C.(ed.): Denitrification, nitrification and atmospheric nitrous oxide, pp. 67- 84, Wiley, New York, Chicester, Brisbane, Toronto.

13. Shapleigh, J. P. and Payne, W. J. (1985). FEMS Microbiol. Letters 26, 275-279.

14. Lam, Y. and Nicholas, D.J.D. (1969). Biochim. Biophys. Acta 180, 459-472.

15. Newton, N. (1969). Biochim. Biophys. Acta 185, 316- 331.

16. Zumft, W.G. and Vega, J. M. (1969). Biochim. Biophys. Acta 548, 484-499.

17. Kim, C.-H. and Hollocher, T.C. (1983). J. Biol. Chem. 258, 4861-4863.

18. Zumft, W.G. and Matsubara, T. (1982). FEBS Letters 148, 107-112.

19. Zumft, W.G. Coyle, C. L., and Frunzke, K. (1985). FEBS Letters 183, 240-244.

20. Snyder, S.W. and Hollocher T.C. (1984).Biochem. Biophys. Res. Comm. 119, 588-592.

21. Long, C. (ed.).(1961). Biochemists' Handbook, p. 96,Van Nostrand, Princeton, Toronto, New York, London.

AN "IN VITRO" ASSAY FOR EVALUATING THE AZOSPIRILLUM WHEAT ASSOCIATION

M. Bazzicalupo*, E. Cresta**, F. Favilli**
*Dipartimento di Biologia Animale e Genetica
**Istituto di Microbiologia Agraria e Tecnica
Università degli Studi di Firenze

Summary: An "in vitro" test to measure nitrogenase activity (C_2H_2 reduction) of A.brasilense, strain 107, str^+ nir^-inoculated on seedlings of wheat (var.Irnerio), is described. The test was also used to determine nitrogenase activity occurring inside and on the surface of the roots, and to count bacteria on the root. Moreover the test we describe could be useful also as a model for the study of the bacterization process. Keywords: Azospirillum, wheat, association, "in vitro" test.

Introduction

The results of ten years of fields and pots Azospirillum bacterization experiments of various Gramineae have been extremely variable from significant negative or zero effects to large highly significant increase in dry matter and total crop nitrogen content (1-2-3). One of the possible causes of this variability could be the lack of adeguate tests for the selection of the associated plant and bacterial genotypes (1).
Improvement on Azospirillum bacterization benefits can be obtained with better knowledge of the specificity of partners since this may have significant implications in establishing efficient nitrogen-fixing associations.
The sole experimental data on specificity are those of Baldani and Dobereiner (4), Dobereiner and De Polli (5) who reported

Azospirillum III: Genetics, Physiology, Ecology
Ed. by W. Klingmüller

that *A.lipoferum* preferentially infects C_4 plants and *A.brasilense* preferentially infects C_3 plants. This results, however contrast with the isolation of both species from C_3, C_4 and CAM plants (6 - 7).

Several "in vitro" tests have been proposed to study the interactions between rhizospheric bacteria and *Gramineae* (8-9-10-11). However "in vitro" tests to reveal differences in the effectiveness of the N_2-fixation of the association of bacterial strains and different plant genotypes have been so far described only for *Rhizobium* (12) and in some extent for the association *Azotobacter* - maize (13).

In the present work we describe a simple "in vitro" test to measure nitrogenase activity (C_2H_2-reduction) and roots colonization of the association of *Azospirillum* with wheat seedling. Such a test could be useful to determine specificity of associations and also as a model to study the bacterization process and the factors affecting the roots colonization during the first stages of plant growth.

Materials and methods

Organisms: Bacterial strain was *A.brasilense* 107 str^+, nir^- isolated by J.Dobereiner (EMBRAPA - Brazil). Wheat was "Triticum vulgare" var. Irnerio a widely used variety from Italy.

Description of the test: Seeds were sterilized with a 4% hypochlorite solution and allowed to germinate on wet filter paper at 30°C for 24 hours. Seedlings were transferred in sterile 200 x 22 mm test tubes containing 10 ml of a mineral solution (MS) (K_2HPO_4 1,37 g; KH_2PO_4 1,05 g; $MgSO_4$ 0,19; Na_2SO_4 0,1 g; $CaCl_2$ 0,01 g; $FeCl_3$ 0,01 g; Na_2MoO_4 0,002 g; H_2O 1000 ml; pH7) and incubated at 30°C in an illuminated room. Seeds were kept dry out of the medium in which roots were soaked. After three days a bacterial washed suspension was inoculated at about 10^5 cells/ml and the atmosphere of the tubes was changed with N_2 plus 0,8 % of O_2. C_2H_2 was injected at a final concentration of 10% after about ten hours. C_2H_4 production was measured by a Pakard gas-chromatograph.

To some of the tubes, referred in results as washed seedlings, C_2H_2 was not added and two days after inoculation the liquid medium was discarded, the tube and the plant were washed twice with sterile water, and air was replaced with N_2 + 0,8 % O_2; C_2H_2 was then injected and C_2H_4 production was recorded as described.

Bacterial counts: Bacteria from the medium were counted by plating appropriate dlution on MSP (14). Bacteria on the roots were counted as follow: pieces of roots were measured and washed twice in sterile water, crushed on mortar, diluted and plated on MSP.

Results and discussion

In order to find the optimal conditions for an association "in vitro" test, we tryed different media as soil, sand and agar, looking for reproducible nitrogenase activity values in each experiment and for controls (bacteria without plant) with zero activity. Our data (not reported) clearly indicated that liquid medium (MS), described in methods, without carbon sources and yeast extract, was the most suitable.

In Table 1 are reported results obtained in one of the experiments carried out as described in methods. Data indicate that almost all the plants were able to support A.brasilense 107 N_2-fixation at an average rate (20,7 nmol · h^{-1} · $tube^{-1}$) close to

Table 1. C_2H_2 reduction assay of intact wheat seedlings inoculated with A.brasilense 107

Number of samples	Active samples	Nitrogenase activity* (nmol of reduced C_2H_2 h^{-1} $tube^{-1}$)		
		Average Value	Distribution	
			Range of activity	N°of samples
			0 - 8	4 (10 %)
40	39 (97,5%)	20,7	8 - 34	32 (80 %)
			34 - 80	4 (10 %)

(*) Data are obtained during 24 hours after C_2H_2 injection

Table 2. C_2H_2 reduction assay of washed wheat seedlings inoculated with A.brasilense 107

Number of samples	Active samples	Nitrogenase activity * (nmol of reduced C_2H_2 h^{-1} $tube^{-1}$)		
		Average value	Distribution	
			Range of activity	N°of samples
			0 - 1	2 (18 %)
11	9 (82 %)	2	1 - 3,5	9 (82 %)

the activity of the same strain in the same conditions with 1 g/l of Na succinate instead of the plant. The majority of the samples (80%) show similar C_2H_2 reduction activities (between 8 to 34 nmol · h^{-1} · $tube^{-1}$) suggesting a high level of reproducibility. The occurrence of few samples with activities far from the average value could be explained by genetic differences among Irnerio seeds.

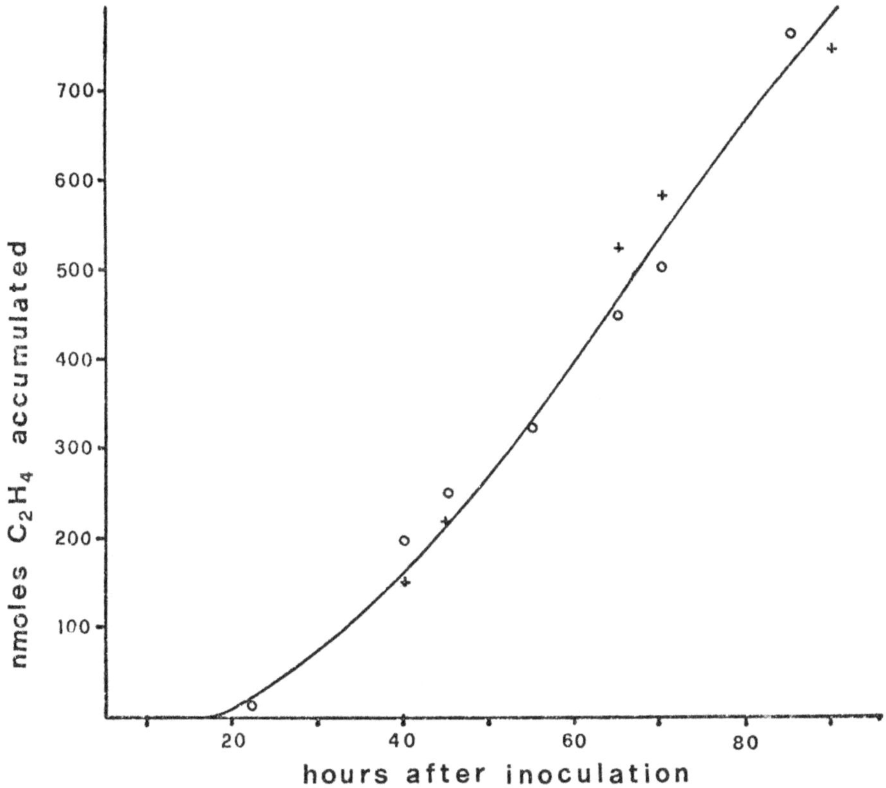

Figure 1. Time course of C_2H_4 accumulation by A.brasilense 107 in the presence of wheat seedling (o), or 0,1 % of Na_2 succinate (x).

Table 2 shows nitrogenase activity of washed wheat seedlings, assayed as described in Methods. Data indicate that 2 days after inoculation N_2-fixing cells of Azospirillum are already linked to the roots. The average C_2H_2 reduction value, that seems highly reproducible, could be considered an important factor for the screening evaluation. The nitrogenase activity was detectable about 4 hours after the washing of seedlings at a constant rate for about 18 hours.

In Fig.1 is reported the time course of nitrogenase activity of Azospirillum supported by a wheat seedling or by 1 g/l of Na suc-

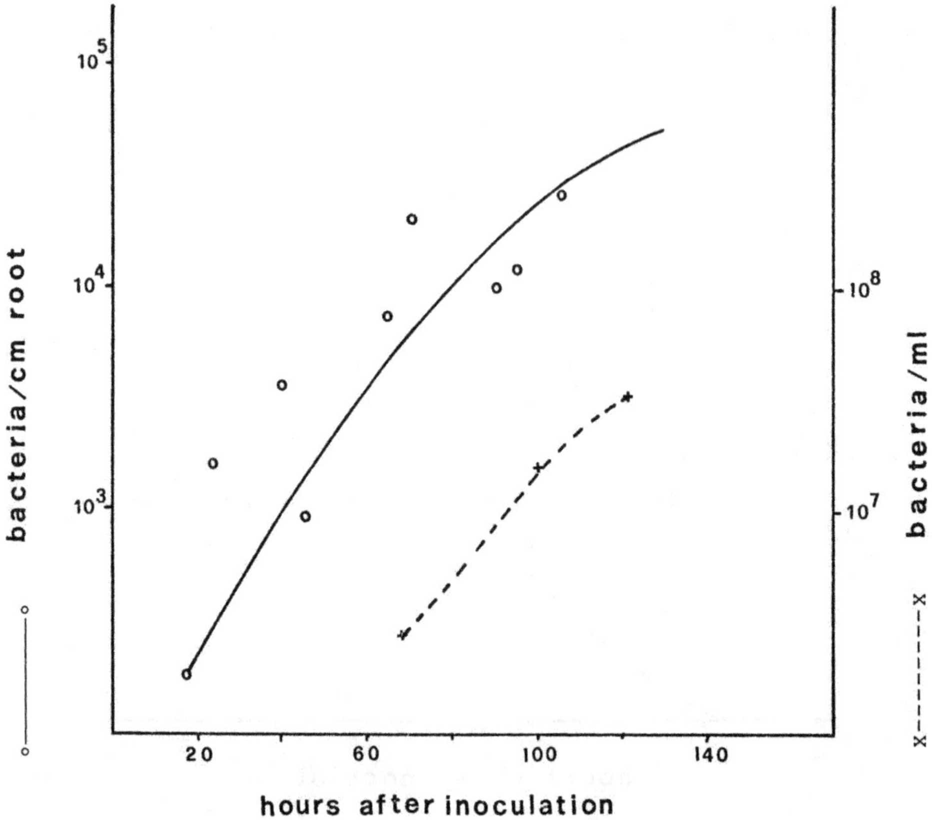

Figure 2. A.brasilense 107 growth during the "in vitro" association experiments: o———o C.F.U. from washed roots; x------x C.F.U. from medium.

cinate. Data obtained indicated the possibility to draw a single curve for both experiments suggesting a high nutritional affinity of the partners, wheat var.Irnerio - A.brasilense 107.
The latter result, moreover, indicate that this kind of experiments could be used for studying factors and conditions which affect nitrogenase activity of associated organisms in a different way than that of bacteria without plant.
During the experiments Azospirillum grew both on the roots and in the medium as showed in the curves of Fig.2.

Because the weight of 1 cm of root is about 2 mg and assuming that the density of roots is similar to that of water, the concentration of bacteria on the root is almost the same of that in the medium; this means that bacteria did not accumulate on the roots. The nitrogenase activity of washed roots was about 10 times lower than the activity of the complete system, as shown in Tab.1 and 2; on the other hand the number of bacteria on the roots of one seedling was at least 100 times lower than total number of bacteria of a tube (the roots of a seedling was 20-40 cm in lenght); this disagreement could be explained by a higher specific activity of nitrogenase of root associated bacteria; further studies will be necessary to better investigate this results.

References

1. Boddey, E.M., Döbereiner J. (1982).XXII Congress of Soil Science, New Delhi, India, p.28.
2. Kapulnik, Y., Okon,Y. (1983) - In: Azospirillum workshop II. Experientia Supplementum, Vol.48, p.163. Klingmuller, W. (Editor), Birkhauser Verlag, Basel.
3. Favilli, F., Balloni, W., Messini, A. (1984) - In: Colloquio sulla Biotecnologia del suolo. Firenze, 11 dicembre 1984 (in press).
4. Baldani, V.L.D., Döbereiner,J. (1980). Soil Biol.Biochem. 12, 433.
5. Döbereiner, J., De-Polli, H. (1980) - In: Nitrogen fixation. Stewart, W.D.P. and Galon, J.R., Editors. Academic Press, London, p.301.
6. Favilli, F., Balloni, W., Cresta, E., Messini, A. (1984) - In: Advances in nitrogen fixation research. Veeger, C., Newton, W.E. (Editors) p.339. Martinus Nijhoff, Junk,W. Publishers.

7. Rao, A.V., Venkateswarlu, B. (1982). Canad.J.Microbiol. 28, 778.
8. Bennett, R.A., Lynch, J.M. (1981) - Journal of Gen.Microbiol. 125, 95.
9. Scher, F.M., Ziegle, J.S., Kloepper, J.W. (1984). Canad.J.Microbiol, 30, 151.
10. Thomas-Bauzon, D., Weinhard, P., Villecourt, P., Balandreau, J. (1982). Canad.J.Microbiol. 28, 922.
11. van Berkum, P., Sloger, C. (1982). Plant Physiol. 69, 1161.
12. Wacek, T.J., Brill, W.J. (1976). Crop Science, 16, 520.
13. Ela, S.W., Anderson, M.A., Brill,W.J. (1982). Plant Physiol. 70, 1564.
14. Bani,D., Barberio, C., Bazzicalupo, M., Favilli, F., Gallori, E., Polsinelli, M. (1980). Journ.Gen.Microbiol., 119, 239.

SECRETION AND CLEAVAGE OF SUCROSE BY WHEAT: A SYSTEM OF CHEMOTACTIC ATTRACTION AND NITROGENASE INDUCTION IN AZOSPIRILLUM LIPOFERUM

D. Hess, D. Heinrich, J. Kemmner and S. Jekel
Institut für Pflanzenphysiologie, Universität Hohenheim
Emil-Wolff-Str. 25, D-7000 Stuttgart 70, FRG

Summary: The interactions between spring wheat and *Azospirillum lipoferum* were studied under test tube conditions. Sucrose was secreted by wheat roots and cleaved by an invertase, which had been released from wheat roots as well. Excretion and subsequent cleavage of sucrose by wheat proved to be a main driving force in associate symbiosis:

1. Sucrose, fructose, and glucose are attracting *Azospirilla* to wheat roots. The chemotactic potential was increased by sucrose cleavage.
2. Fructose, and to a smaller extent glucose, provided energy sources for bacterial growth and dinitrogen fixation.
3. Fructose, and to a smaller extent glucose, induced bacterial nitrogenase activity even on media with nitrogenase repressing levels of N substances. Nitrogenase induction involved bacterial gene activation. Keywords: Sucrose excretion, sucrose cleavage, chemotactic attraction, nitrogenase induction, carbohydrate utilization, association wheat - *Azospirillum*.

Introduction

For ecological and economical reasons, the impact of associative dinitrogen fixation on technical N fertilizer application is under investigation. Since several years, we are studying associations of spring wheat (*Triticum aestivum*) and *Azospirillum lipoferum* under greenhouse and field conditions. In none of our experiments we failed to get a positive res-

Azospirillum III: Genetics, Physiology, Ecology
Ed. by W. Klingmüller

ponse of wheat to bacterial inoculations. The degree of this response, however, varied greatly, in the case of grain yield increase from 8 - 32% (1). Apparently, more insight into the interactions between cereals and bacteria will be needed to optimize associative symbioses, and to avoid failures.

To facilitate the basic studies needed, we used *in vitro* systems. On agar media, wheat roots attracted *A. lipoferum*. Thereafter, nitrogenase activity was induced within the bacteria even on media with repressing levels of N substances (2, 3). A special transfilter system allowed to collect conditioned media not only from wheat or bacteria, but also from associations (4, 5). Investigations using such conditioned media revealed excretion and subsequent cleavage of sucrose by wheat to be a main driving force in attraction of the bacteria, bacterial growth, and nitrogenase induction.

Results and discussion

1. Excretion and cleavage of sucrose

In wheat root exudates high portions of organic compounds, especially sugars, organic acids, and amino acids had been found (recent review: 6). Wheat plants kept in our culture systems excreted mainly carbohydrates, and only minor amounts of amino acids and other organic acids. After 14 days of cultivation in mineral medium, wheat plants had excreted 2.76% of their dry weight in form of sucrose, fructose, and glucose, with sucrose being the main compound in the root exudates.

Additionally, wheat roots released invertase. The sucrose present in the root exudates was cleaved by the enzyme (Fig. 1). Sucrose, which had been supplied to the association medium used by us, was degraded as well (Fig.2). These findings are consistent with an earlier hypothesis, stating a possible cleavage of sucrose by plant invertase released from the roots, which was derived from investigations on associations of pasture grasses with *A. lipoferum* (7).

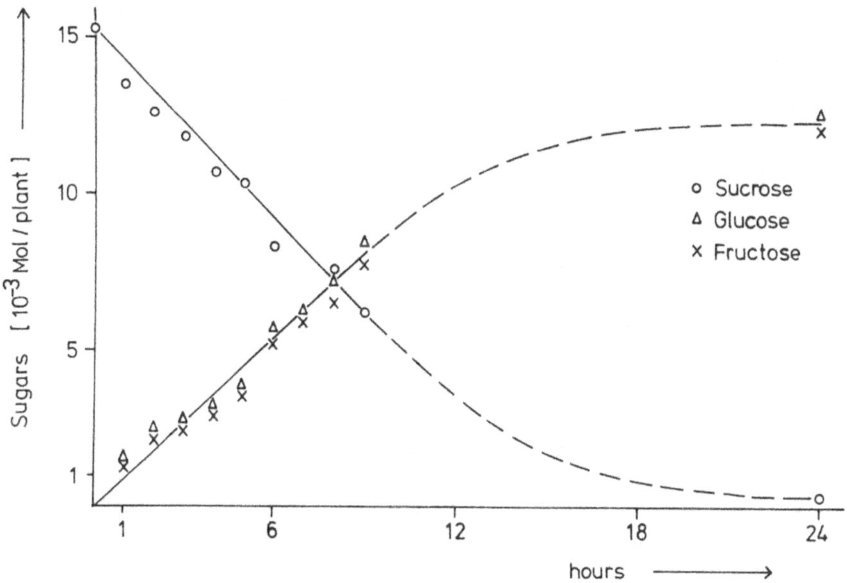

Fig.1. Cleavage of 14C - labelled sucrose by extracellular wheat root invertase. One single wheat plant was grown in mineral medium supplied with uniformely 14C - labelles sucrose At the time intervals indicated, labelled sucrose, glucose, and fructose were separated by thin - layer chromatography and quantitatively determined (from 5).

2. Chemotactic attraction

Using capillary tests (8, 9), chemotactic activity of wheat conditioned media and media from wheat - *A. lipoferum* associations could be demonstrated. Immobilization of the bacteria within the capillaries, pH effects, and an aerotactic response could be excluded (5, 10). Activity of wheat conditioned media was higher than of association media (Fig.3). Both amino acids as well as sugars attract *A. lipoferum*. In root exudates, however, there were only trace amounts of amino acids compared with considerable concentrations of sucrose, fructose, and *A. lipoferum* cannot metabolize sucrose (BM). Wheat plants are

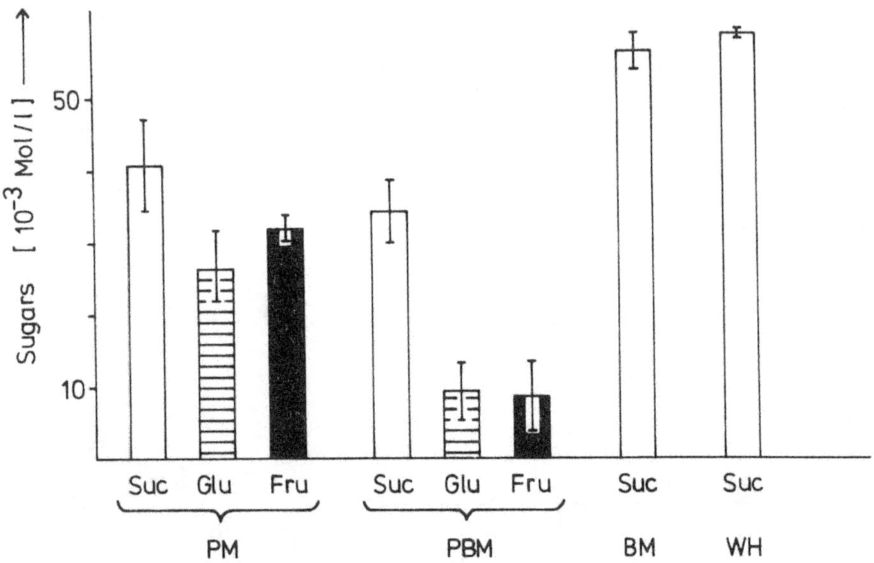

Fig.2. Sugar concentrations in the media used for chemotaxis assays, and nitrogenase induction at the end of the association period (17 days). Su = sucrose, Fru = fructose, Glu = glucose.

PM = plant medium; medium from wheat plants alone.

PBM = plant bacteria medium; medium from wheat - Azospirillum associations.

BM = bacteria medium; medium from Azospirilla alone.

WH = association medium (3).

cleaving the sucrose supplied in the association medium (PM). In associations, the bacteria are metabolizing the glucose and fructose delivered by cleavage of sucrose (PBM) and glucose. Furthermore, in associations on different media the time course of the combined accumulation of sucrose, fructose, and glucose paralleled the time course of chemotactic activity (Fig.4). Therefore, at least in our systems the sugars mentioned are the main attractants. This allows to explain why media from associations showed a lower chemotactic

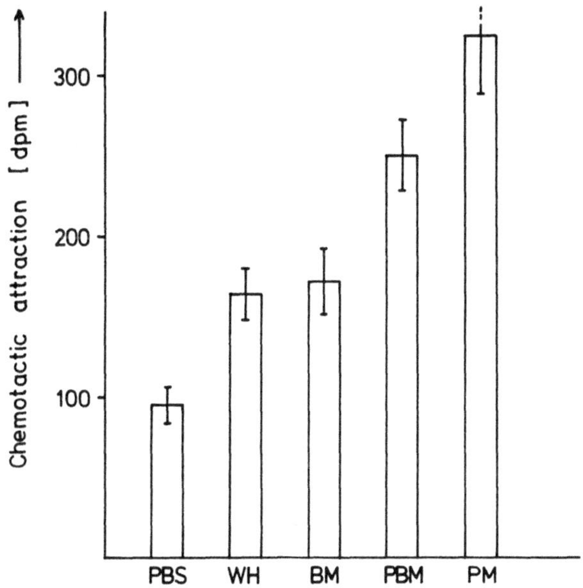

Fig.3. Chemotactic attraction of A. lipoferum by conditioned media. PBS = buffer used for suspending the bacteria (9), other abbreviations as in Fig.2.

actitity than media from wheat alone: part of the attractants, fructose and glucose, are metabolized by the bacteria.

Chemotactic attraction by one unit fructose plus one unit gluocse proved to be significant higher than by one unit sucrose (Fig.5). Therefore, the chemotactic potential was increased by the invertase cleavage of sucrose- a first effect of the sucrose excretion and cleavage system (5, 10).

3. Induction of nitrogenase activity

On agar media (2, 3) and in transfilter associations (4; for the assay system cf. 5) wheat induced nitrogenase activity in A. lipoferum even in the presence of repressing levels of N substances. From transfilter systems conditioned media were

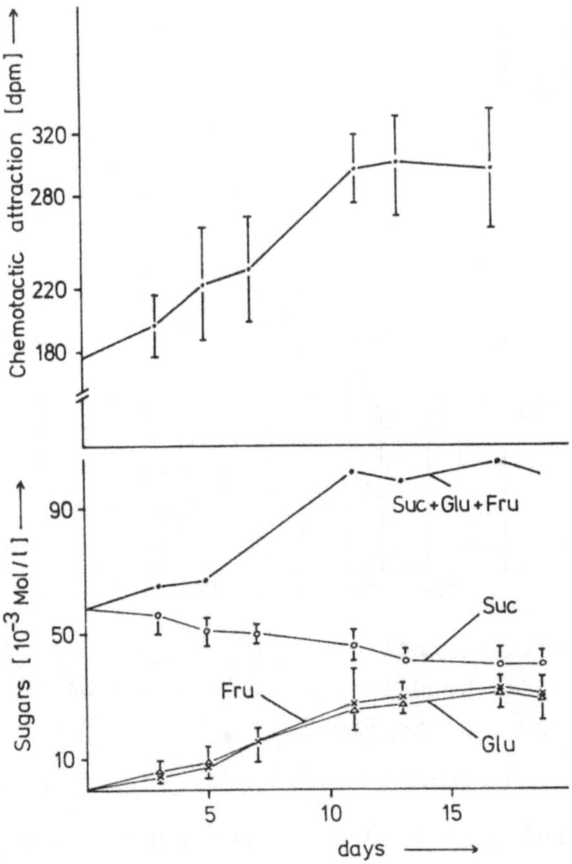

Fig.4. Time course of chemotactic attraction and of sugar concentrations in association medium (WH medium) on which wheat was grown (from 5).

collected and studied further. Measurements of the N content of the media did not support the assumption, that the induction was due to the removal of nitrogenase repressing N substances by the growing wheat plants. Additionally, we assayed concentration and dilution ranges of the respective media. Supposing that repressing N substances were involved, one should expect a concentrating of the media,and,therefore,

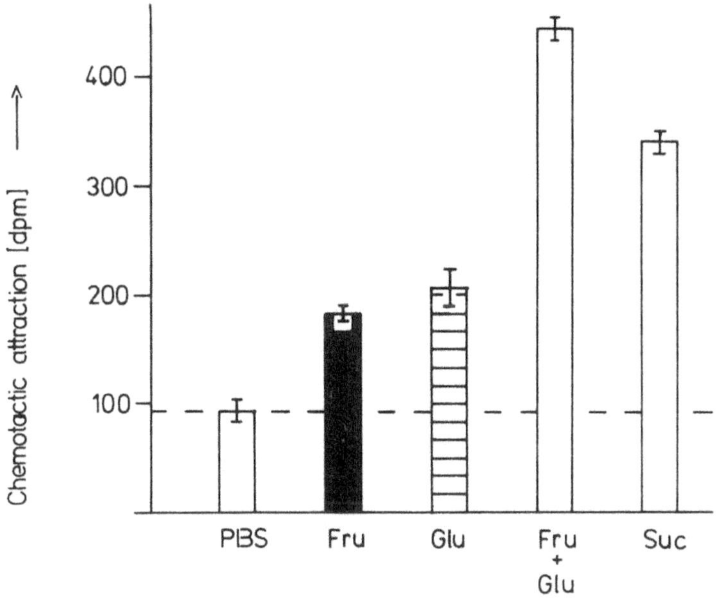

Fig.5. Chemotactic activity of fructose (Fru), glucose (Glu), sucrose (Suc), and an equimolar mixture of fructose and glucose (Fru + Glu). Sugar concentrations 10^{-3} M each. PBS = buffer used for suspending the bacteria and dissolving the sugars.

of repressing substances to cause a decrease in nitrogenase activity. The reverse was true: concentrating the media the inducing capability increased. Therefore, an induction of nitrogenase activity by removal of repressing substances could be excluded.

Usually, media from wheat alone showed an higher inducing activity than media from associations (Fig. 6). This resembled the situation we had found in studying the chemotactic activity of conditioned media, where part of the attractants, fructose and glucose, had been metabolized by the bacteria. Indeed, both fructose and glucose induced nitrogenase activity on media with repressing amounts of N substances. There

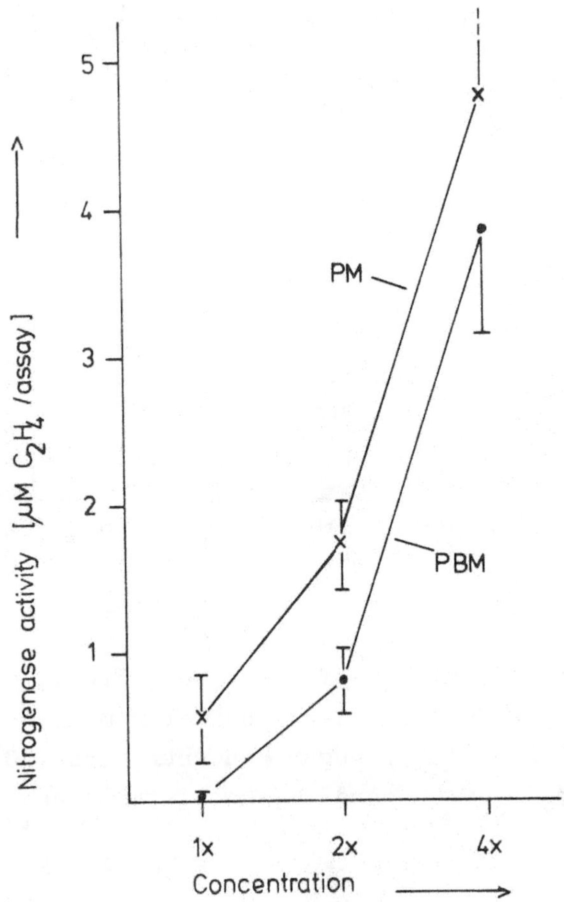

<u>Fig.6.</u> Nitrogenase activity (AR), depending on the concentration of conditioned media. PM and PBM as in Fig.2. The media were concentrated by lyophilization.

were, however, striking differences between both sugars: fructose induced a high level of nitrogenase activity within the first days, whereas the action of glucose was later and weaker (Fig.7). Sucrose does not induce nitrogenase activity.

On mineral medium, the sucrose excreted by the roots was cleaved into fructose and glucose. On association medium with

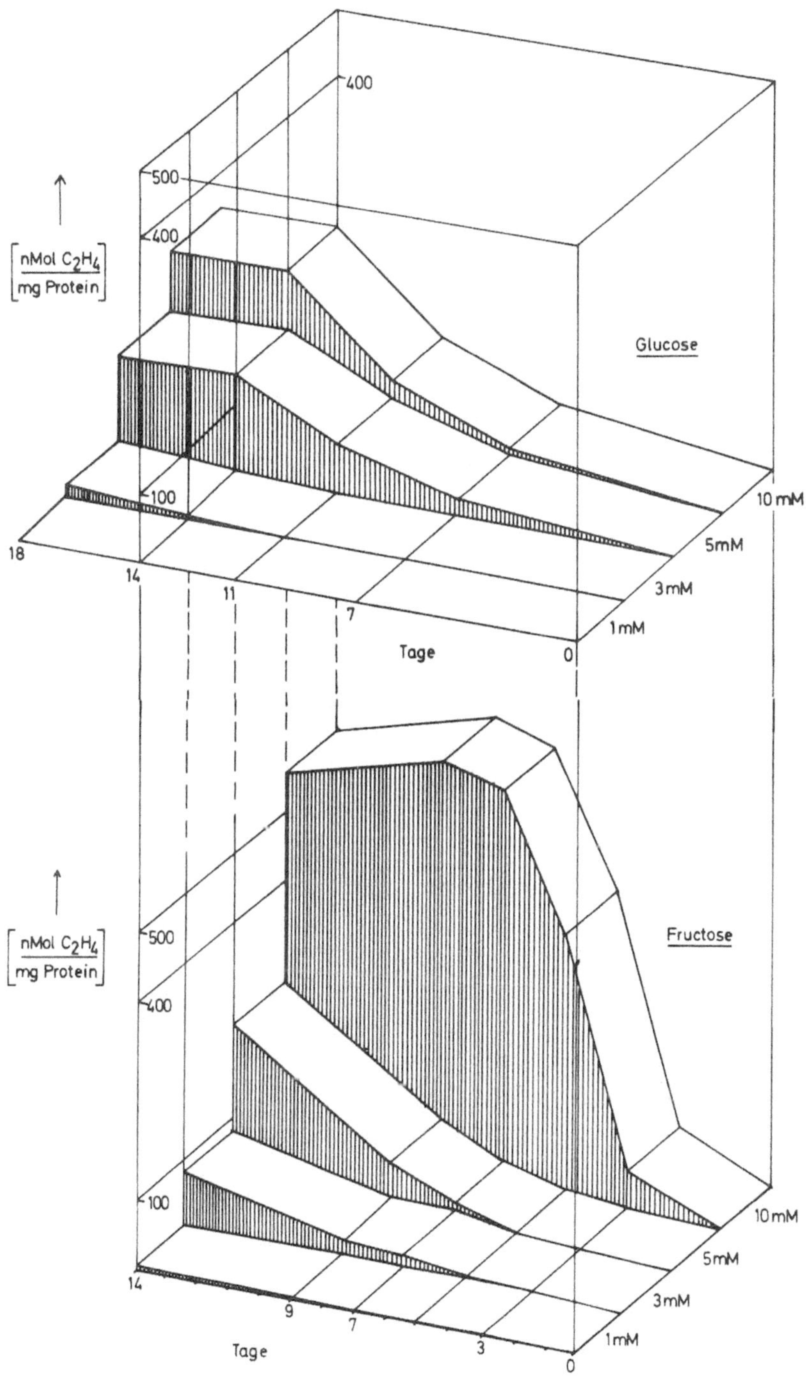

Fig.7. Time course of nitrogenase induction (AR) in Azospirillum lipoferum by different molarities of fructose and glucose. Tage = days.

its high sucrose content (2%; 3) cleavage of the sucrose supplied with the medium was more important. As expected, fructose and glucose were always found in a 1 : 1 ratio (Fig. 2). The time course of the combined accumulation of fructose and glucose, and of fructose alone paralleled the development of nitrogenase activity in associations, indicating that these sugars, first of all fructose, are the main inducing factors. Therefore, the second effect of the sucrose excretion and cleavage system will be the delivery of nitrogenase inducing compounds, especially fructose (11).

4. Bacterial growth and sugar metabolism

In further investigations we tried to find out, why fructose and glucose showed differences in inducing nitrogenase activity. One possibility was, that they were not equally metabolized by the bacteria.

The medium used for wheat - *A. lipoferum* associations was highly supplied with organic compounds including sugars (2% sucrose, 0.1% glucose). It allowed normal growth of wheat as well as of *A. lipoferum* (3). Due to its complex composition, however, it was not suitable to get first insight into the differential action of fructose and glucose.

Therefore, we replaced the association medium by PBS (9), and added the sugars under study. As expected, bacterial growth was greatly reduced compared with association medium. On sucrose, there was no bacterial growth at all. On glucose we observed a very limited, on fructose a better, and on equimolar mixtures of glucose and fructose an even better growth (Fig.8).

This will be the third effect of the sucrose excretion and cleavage system: to provide energy sources for bacterial growth and dinitrogen fixation. In addition, the above data demonstrate, as expected from the induction experiments, dif-

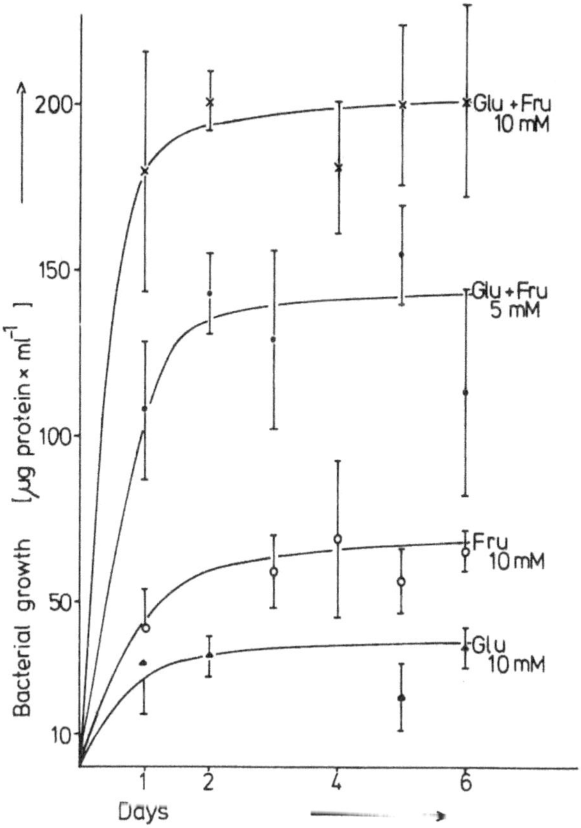

Fig.8. Growth of Azospirillum lipoferum on PBS plus fructose, glucose, and equimolar mixtures of fructose and glucose, respectively. In the mixtures, the concentration of each sugar was 5 mM and 10 mM, respectively.

ferences between fructose and glucose. In supporting bacterial growth just as in nitrogenase induction, fructose proved to be much more efficient than glucose.

Corresponding data were obtained in studying the exhaustion of fructose and glucose from PBS by Azospirilla. If fructose and glucose were provided separately, 10 mM fructose was completely exhausted from PBS within 6 days, whereas glucose was only slightly used (Fig.9). If both sugars were given together, fructose was exhausted at the same rate as before, whereas glucose utilization was completely prevented (Fig.10). Once more a preference for fructose became apparent.

The mechanism underlying these effects remains to be elucidated. Investigations are initiated to clarify, whether we are dealing with an uptake phenomenon, or with metabolism _in sensu strictu_.

5. Induction of bacterial gene activity

Induction of nitrogenase activity was prevented by rifampicin. The concentration of the antibiotic used in these experiments did neither impair with wheat development nor with bacterial growth. Therefore, unspecific toxic effects could be excluded.

Rifampicin counteracted plant conditioned media as well as chemically defined components from these media. Fig. 10 demonstrates nitrogenase induction by fructose and glucose, respectively, with and without rifampicin added.

It seems highly probable, that the bacterial gene material activated by glucose and much more by fructose comprised the nif operon. To obtain certainty, "Northern" blotting studies are initiated.

The molecular mechanism of gene activation remains to be elucidated. Enhancement of metabolism might well be followed unspecifically by gene activity.

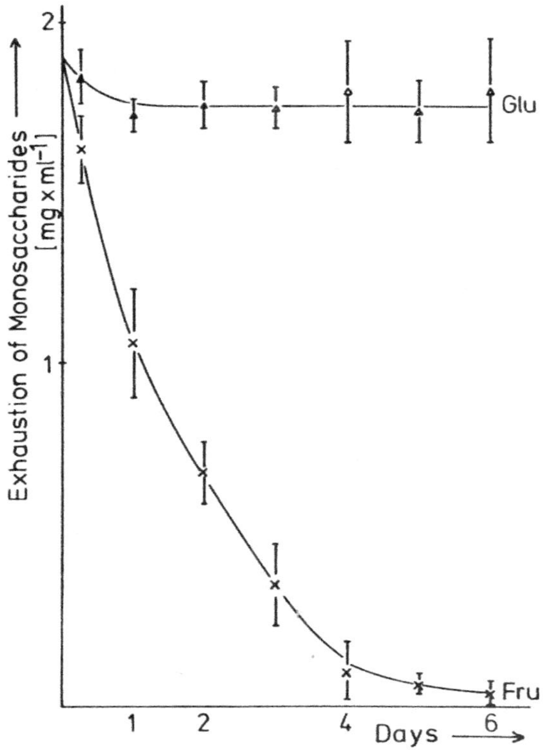

Fig.9. Exhaustion of fructose and glucose from PBS by A. lipoferum. Fructose and glucose were given in separate experiments. Sugar concentrations at time 0: 10 mM each. The data obtained correspond with the growth behavior of A. lipoferum on fructose and glucose, respectively (cf. Fig.8).

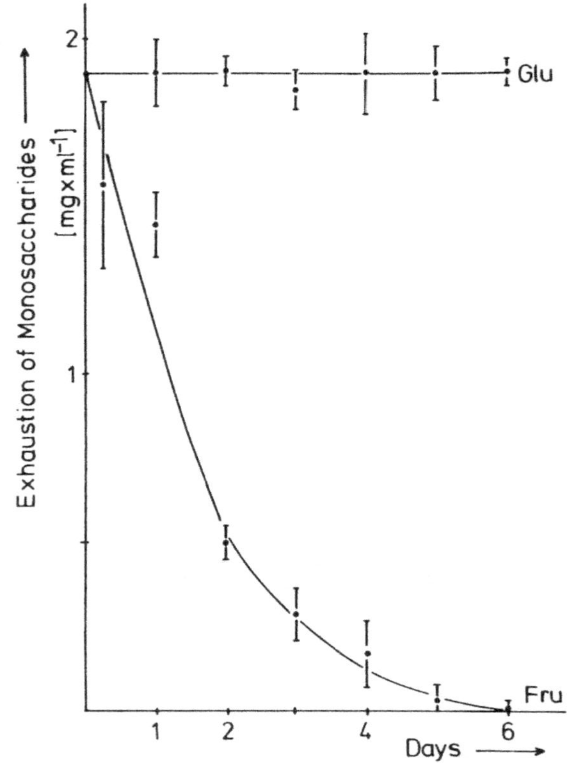

Fig.10. Exhaustion of fructose and glucose from PBS by A. lipoferum. Fructose and glucose were given together (cf. Fig.9). Sugar concentrations at time 0: 10 mM each. The data obtained correspond with the growth behavior of A. lipoferum on equimolar mixtures of fructose and glucose (cf. Fig.8).

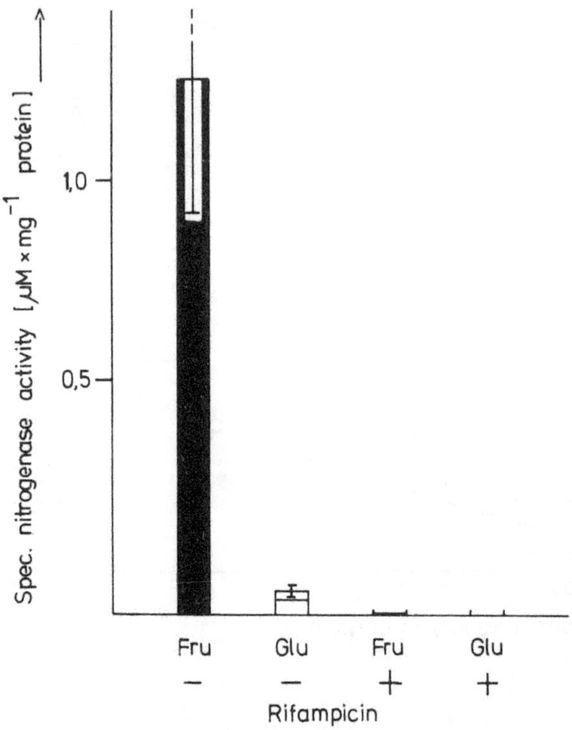

Fig.11. Nitrogenase induction in *A. lipoferum* by fructose and glucose, respectively, with and without rifampicin (0.01 mg x ml^{-1}) added.

Conclusions

In wheat - *A. lipoferum* associations, sucrose, a commonly available plant carbohydrate, proved to be the starting point of a sequence of events. First, sucrose excreted by the wheat roots was cleaved into fructose and glucose by an invertase, which had been released from the roots as well. This sucrose excretion and cleavage has consequences:

1. The chemotactic potential of the roots was increased.
2. Energy sources for bacterial growth and dinitrogen fixation were provided.

3. Nitrogenase was induced, probably by activation of the nif operon. Induction occured even on media with nitrogenase repressing levels of N substances.

In bacterial metabolism, bacterial growth, and nitrogenase induction, sucrose was ineffective. Considering its cleavage products, fructose proved to be much more efficient than glucose.

Taking into consideration data which had obtained in associations of pasture grasses with *A. lipoferum*, it seems possible, that this sucrose excretion and cleavage system will be no pecularity of associations with wheat, but of general occurance.

As mentioned, fructose induced nitrogenase activity in *A. lipoferum* cultivated on repressing levels of N substances. Therefore, on could speculate about its possible application under field conditions even if medium levels of N fertilizers had been added. Fructose, however, seems to be expensive. Therefore, we are screening organic compounds, especially fructose derivatives, which might replace fructose. Until now, we were unable to find a better nitrogenase inducer in *A. lipoferum* than fructose.

Acknowledgements

Part of the work was financially supported by the Deutsche Forschungsgemeinschaft and especially by the Vater und Sohn Eiselen Stiftung, Ulm. Dr. W. Wenzel supervised the English.

References

1. Mertens, Th. and Hess, D. (1984). Plant Soil 82, 87 - 99
2. Hess,D. (1980). Inf.Landwirtschaftsberatung Baden - Württemberg 5, 63 - 80

3. Hess, D. and Kiefer, S. (1981). Z.Pflanzenphysiol.101, 15 - 24

4. Hess, D. (1982). Experientia Suppl.42, 69 - 74

5. Heinrich, D. and Hess, D. (1985). Can.J.Microbiol. 31, 26 - 31

6. Rovira, A.A., Bowen, G.D. and Foster, R.C. (1983). In Enyclopedia of Plant Physiology. New Series Vol.15A. Edited by A.Läuchli and R.L.Bieleski. Springer, Berlin. pp 61 - 93

7. Martin,P. and Glatzle,A. (1982). Experientia Suppl.42, 108 - 120

8. Adler,J., Hazelbauer,G.L. and Dahl,M.M. (1973). J. Bacteriol. 115, 824 - 847

9. Currier, W.W. and Strobel,G.A. (1976). Plant Physiol.57, 820 - 823

10. Heinrich, S. (1984). Dr. rer.nat. Thesis Universität Hohenheim

11. Hess, D. et al. in preparation

FIELD INOCULATION OF RICE USING IN VITRO SELECTED BACTERIAL AND PLANT GENOTYPES

P.B.B.N. Charyulu[1], F. Fourcassie[2], A.K. Barbouche[3]
L. Rondro Harisoa[4], A.M.N. Omar[5], P. Weinhard[6]
R. Marie[7] and J. Balandreau[8]

1. 6. and 8. Centre de Pédologie Biologique du C.N.R.S., B.P. 5, 54501 Vandoeuvre-les-Nancy Cédex, France.
2. Azote et Fertilisants (AzF), Centre de Recherche Agronomique, 132, route d'Espagne, 31057 Toulouse Cédex, France.
3. Ecole Supérieure d'Agriculture; Le Kef, 7119 Tunisie.
4. CALA-FOFIFA, B.P. 30, Station Alaotra, Madagascar.
5. Unit of Biological Nitrogen Fixation, Soils and Water Research Institute, Agriculture Research Center, Giza, Egypt.
7. INRA, Station d'Amélioration des Plantes, 9, place Viala, 34060 Montpellier Cédex, France.

Summary : A collection of N_2 fixing bacterial strains was isolated from the rhizosphere of rice. Under gnotobiotic conditions they were associated with rice plantlets and their acetylene reduction activities compared. The most active was strain 4B of *Azospirillum lipoferum*. In a similar manner, this strain was associated with a collection of rice mutants of cv Césariot. The most active mutant was nº 17. Three field experiments were performed associating strain 4B with Césariot or its mutant nº 17. Positive effects on yield or decreases in N natural abundances were obtained following inoculation except under conditions of restricted aeration of roots.... Keywords : *Azospirillum*, rice, inoculation.

Introduction

In previous papers (9, 15), we have described a new enrichment procedure using germinating seedlings for the isolation of rhizosphere dominant N_2-fixing bacteria. This gnotobiotic model called

Azospirillum III: Genetics, Physiology, Ecology
Ed. by W. Klingmüller

"spermosphere model" has been used to compare nitrogenase activities of motile and non motile clones of Azospirillum lipoferum (9, 10). Considering the current lack of available criteria for choosing bacterial and plant genotypes in field inoculation studies, we further investigated the possibility to generalize the use of the spermosphere model to select optimal combinations of rice and bacterial genotypes in the laboratory before field inoculation trials.

Materials and methods

1. Plants : rice cultivars (Arlésienne, Césariot, Cigalon, Cristal, Delta) were obtained from INRA, as well as a series of mutants of cv Césariot generated by γ-irradiation of seeds (11, 12); cv IR8 was obtained from IRRI. Seeds had been collected in the same year and stored under the same conditions.

2. Bacteria : representative N_2-fixing bacteria were isolated from a french rice-field rhizosphere soil using the technique described previously (15). They were determined by BALLY et al. (4) : 2 AB1 is a Klebsiella oxytoca, 7 ATr is an Enterobacter cloacae, 4B and 4T are Azospirillum lipoferum (4T is non motile) and 5 ATr a "Pseudomonas" paucimobilis (8) ; all of them had natural abundances higher than 10^5 per g of dry rhizosphere soil. Strain B8 is a Beijerinckia sp. isolated using more conventional methods from another rice field in France. Strain 7001 is a Streptomycin resistant derivative of Azospirillum brasilense Sp7 ATCC 29145, obtained by C. ELMERICH (Pasteur Institute, Paris).

3. The "spermosphere model" is the name given to our gnotobiotic model (15). It is made of a Pankhurst tube (13) in which a few invaginations can hold the germinating seed at the surface of the nutrient medium (Figure 1). The side arm contains 2 ml of NaOH, 1 N as a CO_2 trap. The tubes are closed by rubber stoppers (Subaseal type) and incubated in the dark.

4. Media : to prepare the inoculum, bacteria were grown in regular nutrient broth (Difco). They were centrifuged and washed

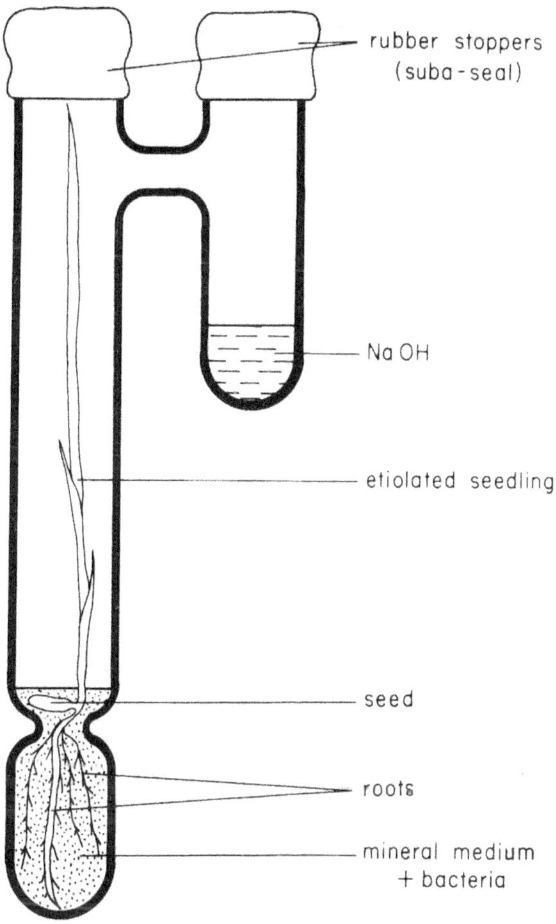

Figure 1. The spermosphere model . It consists of a Pankhurst tube containing 6 ml of modified Weaver's medium, a mineral medium without C and N. The side tube contains sodium hydroxide 1 N to trap CO_2.

twice, then resuspended in modified Weaver medium (16) to a concentration of 10^8 bacteria ml^{-1}. Modified Weaver medium has the following composition :

- Solution A (mgl^{-1}) : Zn SO_4, 7 H_2O, 0.43 ; $MnSO_4$, H_2O, 1.3 ; Na Mo O_4, 2 H_2O, 0.75 ; H_3BO_3, 2.8 ; $CuSO_4$, 7 H_2O, 0.026 ; Co SO_4, 7 H_2O, 0.07.
- Solution B (mgl^{-1}) : Mg SO_4, 7 H_2O, 100 ; $CaCl_2$, 2 H_2O, 100 ; $FeSO_4$, 7 H_2O, 22 ; EDTA, 20 ; Solution A, 20 ml.
- Solution C (gl^{-1}) : KH_2 PO_4, 0.6 ; K_2HPO_4, 0.9.

Add 50 ml of B and 15 ml of C to 1 liter of distilled water. Every tube contains 6 ml of this medium.

5. Sterilisation and inoculation : seeds were decorticated and those with a stain or a crack were discarded. Seeds were then surface sterilized for 20 minutes in 10 % hydrogen peroxide, rinsed in sterile water, then immersed in satured $Ca(ClO)_2$ for 2 h and rinsed again in sterile water. One seed was dropped into each tube which was then closed with Suba-seal stoppers and incubated at 28°C in the dark. When coleoptiles were 1 cm high, the plantlets received 10 % C_2H_2 and were inoculated through the stopper by 1 ml of inoculum containing 10^8 bacteria. Propane was used as an internal standard. Ethylene evolution was monitored for 2 weeks and tubes checked for contamination at the end of the experiment. Contaminated tubes were discarded. The number or remaining replicates was never less than 5 and up to twenty.

6. ^{15}N natural abundances determinations : The rice cultivars and mutants were grown side by side under regular conditions of N fertilization (150 kg N ha^{-1}) in small plots (2x2 m approximately). Seeds were collected from the medium part of mature panicles and mixed. A subsample of 100 seeds was taken, ground and mineralized. N isotopic determinations were performed using a double channel mass spectrometer in the mass spectrometry department of Lyon I University. In the field inoculation experiments, harvested seeds were processed in the same way and isotopic determinations were done either in Lyon or in the "Laboratoire de Géologie Dynamique" of Paris VI University.

7. First field experiment : In this experiment uninoculated Césariot was compared with its mutant n° 17 inoculated by <u>Azospirillum lipoferum</u> 4B. Plantlets were prepared in spermosphere models . M17 only was inoculated. After two weeks, the contents of spermosphere models were trnasferred into peat pots containing a mixture of soil and vermiculite and kept under laboratory conditions ; two weeks later the plantlets were transplanted in the field. Each of the two treatments (M17 + 4B, uninoculated Césariot) was assigned a small plot (0.42 m²) in the middle of 2 large plots planted with the rice cv Delta (which is, itself, a mutant of Césariot). Those two plots were cultivated in the regular way and received 150 kg N per ha (100 kg N as $(NH_4)_2SO_4$ at sowing and 50 kg N as $(NH_4)_2SO_4$ at tillering). The numbers of tillers (vegetative and fertile) were evaluated. Dry matter of seeds and their N contents were measured as well as their ^{15}N abundances. The surrounding cv Delta plants were also sampled and used as controls. For each small plot, ^{15}N abundance was compared with the respective Delta control. Due to the higher ^{15}N content of soils compared to atmospheric N_2, a plant fixing more N_2 is expected to be poorer in ^{15}N (1, 2, 5, 6, 14).

8. Second field experiment : This experiment contained four treatments :

- Césariot inoculated by strain 4B
- Césariot inoculated by autoclaved 4B
- M17 inoculated by 4B
- M17 inoculated by autoclaved 4B

Each treatment was in five replicates. Each of the twenty plots was 2 m wide and 5 m long and received water from one end only ; plots were separated by polyethylene sheets to avoid lateral transfer of bacteria.

Seeds were immersed overnight in a dense suspension of bacteria (killed or alive) before planting in a greenhouse nursery on April 30. Before transplanting (June 27) the plantlets had their roots immersed in a similar bacterial suspension for 18 hours.

Beside P and K fertilizers, the plots received 120 kg N ha^{-1} as ammonium sulphate.

Cold weather at flowering and birds during grain filling caused a large incidence of empty seeds so that the results were better expressed as theoretical yields calculated after the number of seeds per square meter and the mean weight of normal seeds.

9. Third field experiment : This experiment contained eight treatments : two rice genotypes (Césariot, M17), two levels of N (0 and 92 kg N ha^{-1}), inoculated (4B) or not. There was only one replicate per treatment. Each treatment was set in a cement tank containing homogenized soil and burried in the rice field ; tanks were 0.4x0.4x0.4 m and received 16 seeds ; this corresponds to a plant density of 10^6 plants per ha. Each seed received a peat based inoculum containing 10^7 bacteria. The non inoculated controls received only peat.

At harvest the component of yield were determined and the isotopic composition of seed N was measured.

The first and the third field experiments were established in the INRA experimental rice field and the second one was in a farmer's rice field. All three sites are located in Camargue, in the Rhône delta.

Results

1. Spermosphere model

This experimental model is based upon the discovery (7) of an initial phase of nitrogenase activity during the early period of plant development which is not supported by photosynthesis. During this period, exudates originated mainly from the products of hydrolysis of seed storage compounds and nitrogenase activity occurs even in the dark. Several assumptions were made at the beginning of this work :

- Plants with a high N_2 fixation potential have a high rate of exudation.

- The rate of exudation and the exudate composition do not depend upon the origin of C exuded (seed storage products or photosynthesis).

- Bacteria which actively fix N_2 in association with seedlings, and plantlets which actively support N_2 fixation, continue to do so during later stages of plant growth.

Heulin _et al_. (9) have shown that in spermosphere model acetylene reduction proceeds with a maximum lag of one day and is linear for more than 15 days. A plateau follows due to the lack of oxygen : C_2H_2 reduction starts again when O_2 is injected at 25th day.

During this linear phase exuded C is the limiting factor : adding Na malate significantly increases the rate of C_2H_2 reduction (E. Spiquel, personnal communication, data not shown).

Spermosphere models being incubated under 10 % C_2H_2 bacterial growth is N limited and very poor. Accordingly we checked that the number of bacteria was not limiting the reduction of acetylene. Table 1 shows that it is not the case.

Table 1. Ethylene evolution by the association rice cv Delta + _Azospirillum brasilense_ Sp7 in spermosphere model with three different inoculum concentrations. Nanomoles per plant ± confidence intervals (.90).

Time (days)	Number of bacteria inoculated per plant		
	10^6	10^7	10^8
7	1760±160	1880±130	1610±120
13	3460±280	3970±260	3150±260
20	5060±310	5670±430	5190±580

Moreover, in this model, plant to plant variability is very low and can be further decreased by using seeds of the same initial weight and discarding abnormal plantlets. This allowed us to easily evidence differences between bacterial strains (9) or rice genotypes (9). Henceforth, we characterized each bacterial-plant combination by the ethylene rate of evolution expressed as percent of a reference combination : cultivar Césariot inoculated by *Azospirillum lipoferum* 4B.

2. Comparison of bacteria

We first compared six strains representative of different members of the dominant N_2 fixing flora associated with rice in the south of France. For the sake of comparison we included strain 7001, a derivative of the well known Sp7. Bacteria were associated with cv Delta.

Large difference appeared between strains (Figure 2), *Klebsiella* 2AB1 being the least active and *Azospirillum lipoferum* 4B being the most active.

3. Comparison of plant genotypes

The comparison of plant genotypes was made difficult by existing differences in mean seed weights between cultivars and mutants. In a preliminary study of cv Delta we could evidence a significant (.99) linear correlation $y = 0.049\ x - 0.225$ between acetylene reduction (10^{-6} C_2H_4 day^{-1} $plant^{-1}$) and x, the initial decorticated seed weight (mg). Assuming a similar relationship in other cultivars, we corrected acetylene reducing activities to take into account such seed weight differences between plant genotypes. All the results shown in the paper have been made comparable by this calculation and refer to a same initial seed weight of 27 mg.

Differences between plant cultivars are large (Table 2) and still larger between mutants (Figure 3). Mutagenesis produces new genotypes with both higher and lower N_2 fixation supporting activities. The most active mutant of cv Césariot is mutant 17.

To check the validity of this ranking we established another ranking based on seed ^{15}N natural abundances of a few rice genotypes grown in the field without inoculation. Table 2 indicates

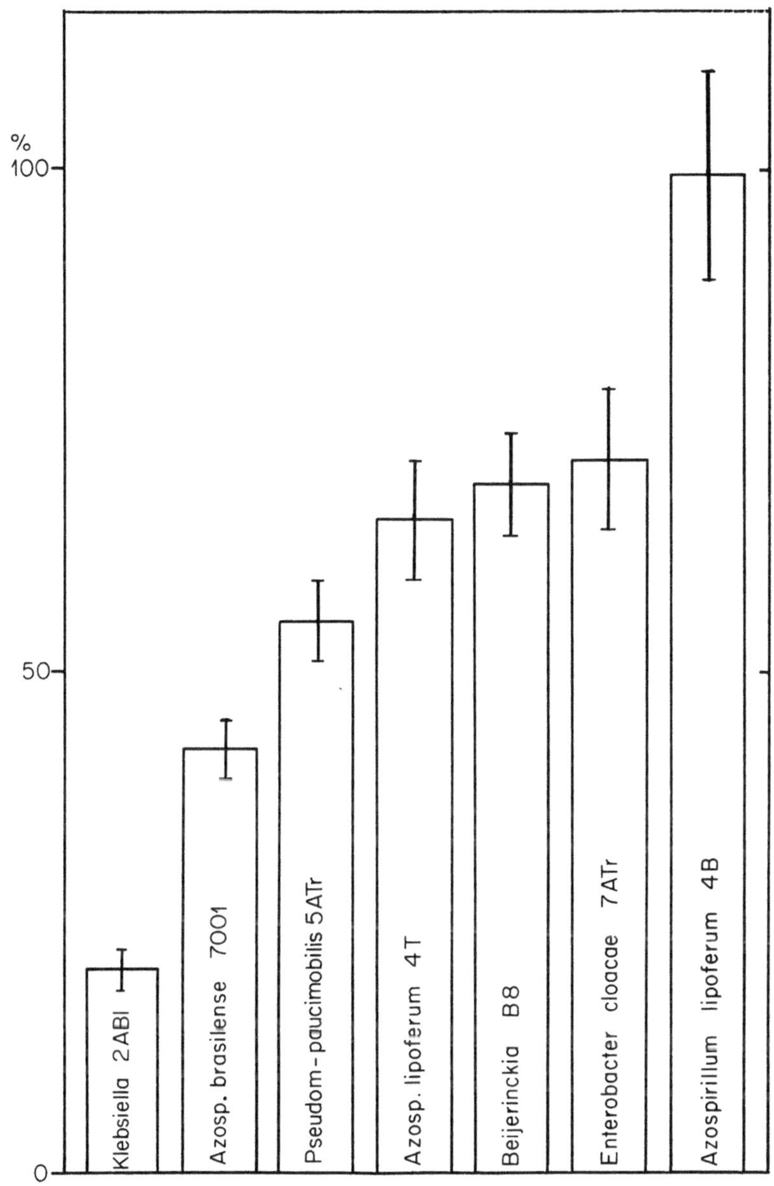

Figure 2. Comparison of bacterial strains in spermosphere model, when associated to cv Delta. Each strain is characterized by its rate of ethylene evolution during the linear phase and expressed as percent of strain 4B rate ± confidence interval (.90).

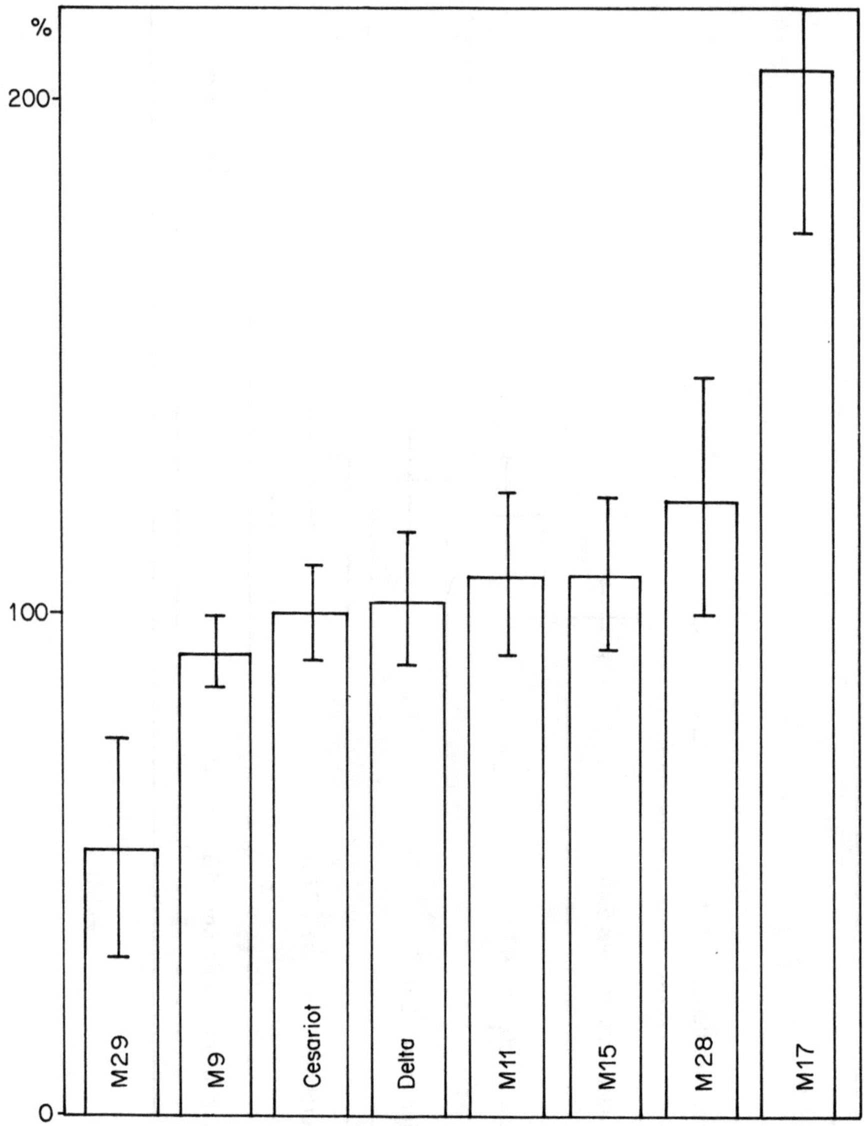

Figure 3. Comparison of rice genotypes in spermosphere model when associated with Azospirillum lipoferum 4B. Each genotype is characterized by the rate of ethylene evolution during the linear phase and expressed as percent of cv Césariot rate ± confidence interval (.90).

that there is some similarity between the two rankings : cultivars with the lowest seed ^{15}N abundances have the maximum nitrogenase activities in spermosphere model. This is indicative of a permanence of the Nis (N_2-fixation support) phenotype from germination to maturity.

Table 2. Comparison of cultivar rankings at germination using acetylene reduction in spermosphere model and at maturity using naturel abundances of ^{15}N. Means ± standard errors. Results are significantly different ($p<0.05$) when followed by differents letters. Student's (spermosphere model) or Wilcoxon's test (% ^{15}N).

Cultivar	Spermosphere model 10^{-9} C_2H_4 day^{-1} $plant^{-1}$	% ^{15}N in mature seeds
IR8	1712±113 a	n.d.
Cigalon	1479±133 ab	0.3705±0.0008 (a)
Césariot	1460± 75 b	0.3706±0.0009 (a)
Delta	1428±100 b	0.3716±0.0009 (b)
Arlésienne	1284± 78 bd	0.3710±0.0006 (b)
Cristal	1167± 53 d	0.3714±0.0006 (b)

From our results it thus appears that M17 is the most active mutant and 4B the most efficient bacterium, at least under the conditions of the spermosphere model. To check this selection procedure it was then necessary to go to the field and establish there the selected association.

4. First field experiment

In a preliminary experiment we compared Césariot grown in the regular way with its mutant 17 inoculated by 4B, in two microplots receiving 150 kg N fertilizer.

Under these conditions (Tables 3 and 4) there is no effect of inoculation on yield but a significant effect on the isotopic composition of seeds : inoculated M17 was lower in ^{15}N than non inoculated Césariot in spite of an opposite difference between the two plots : the seeds of cv Delta plants surrounding inoculated M17 were higher in ^{15}N than seeds of cv Delta plants surrounding control Césariot, reflecting a difference between abundances of ^{15}N in the soils of the two plots.

This result suggests that under high levels of N fertilizers inoculation has still a effect on the origin of plant N : more N is derived from the atmosphere.

Table 3. Results of a preliminary field experiment. Plots were 0.42 m².

	Weight of seeds per plot (g)	N yield per plot (g)	Number of panicles per plot	^{15}N abundances % ±st. dev.
Inoculated M17	166.4	1.87	101	0.3688±0.0002
Surrounding Delta				0.3702±0.0004
Control Césariot	166.8	1.95	122	0.3698±0.0009
Surrounding Delta				0.3695±0.0003

Table 4. Statistical significance of data : ARA (acetylene reducing activities) and ^{15}N natural abundances.

Spermosphere model :	
ARA (M17) > ARA (Césariot)	P = .999
ARA (Delta) = ARA (Césariot)	P = .80
ARA (M17) > ARA (Delta)	P = .95
^{15}N natural abundances :	
% ^{15}N (M17) < % ^{15}N (Césariot)	P > .90
% ^{15}N (Delta) = % ^{15}N (Césariot)(1)	P > .80
% ^{15}N (M17) < % ^{15}N (Delta)(1)	P > .95

(1) M17 and Césariot values are compared with values for respective surrounding Delta.

5. Second field experiment

In this experiment inoculation (Table 5) brought about a large increase in theoretical yields : 700 kg of seeds ha^{-1} in the case of Césariot, 1 000 kg in the case of M17 in spite of a high level of N fertilization (120 kg N ha^{-1}). Observed differences were highly significant (.001)

δ ^{15}N values are not different in the case of Césariot (Table 6). There is a decrease in δ ^{15}N due to inoculation in the case of M17 ($/2 < p < 0.3$) as theoretically predicted.

Table 5. Second field experiment. Theoretical yields expressed as quintals per ha.

	Césariot	M17
Azospirillum	44.6	32.6
Killed Azospirillum	37.6	22.8

Table 6. Second field experiment. Isotopic composition of plants (δ ^{15}N).

	Césariot	M17
Azospirillum	5.42	5.42
Killed Azospirillum	5.46	5.84

6. Third field experiment

The density of plants was 16 per tank, i.e. 16 dm^2 which corresponds to 10^6 plants ha^{-1}. Nets prevented bird problems. Isotopic determinations have been made on one plant per tank ; this plant was selected so that it represents the tank mean.

Table 7. Third field experiment, in tanks, without N fertilizer.

	g paddy per plant	Panicles : number/plant
Césariot		
Inoculated	4.50xxx	3.13xxx
Non inoculated	2.26	1.69
M17		
Inoculated	4.31xx	2.88N.S.
Non inoculated	3.39	2.69

Differences with the corresponding non inoculated control are significant at the following levels :
xxx p $<$.001 ; xx p $<$.01 ; x p$<$.1 ; N.S. p $>$.2

Without addition of N fertilizer (Table 7), there is a strong positive effect of inoculation on yield.

Table 8. Third field experiment ; fertilized tanks (92 kg N ha^{-1})

	g paddy per plant	Panicles : number/plant
Césariot		
Inoculated	3.31x	2.87N.S.
Non inoculated	4.91	3.44
M17		
Inoculated	3.64x	2.50xx
Non inoculated	5.08	3.50

Differences with the corresponding non inoculated control are significant at the following levels :
xxx p $<$.001 ; xx p $<$.01 ; x p $<$.1 ; N.S. p $>$.2

On the contrary, with the addition of N fertilizer (92 kg N ha^{-1}) inoculation has a detrimental effect on yield.

The only possible explanation for this result seems to be denitrification. The strain used is capable of respiring nitrate- through their reduction to N_2 and in this experiment two factors are thought to have favoured this denitrification :

- The water level was not controlled efficiently enough during the growth of rice. Several times submergence water disappeared. This could have caused an increased aeration and nitrification.

- Subsequent readjustments of the water level could have caused drops in redox potential and the more so as the root systems were enclosed in the cement tanks preventing oxygen diffusion from lower, more oxidized layers of soil. These decreases in redox potential following nitrate formation are very likely to stimulate denitrification, especially in the treatments receiving ammonium sulfate ; this could be the origin of the observed detrimental effect of inoculation in the later case.

Conclusion

In many instances, inoculation trials have been done on cereals using bacterial strains from international collections. Little consideration has been given to the origin of these strains, as if N_2-fixing bacteria could accomodate themselves to any soil, to any resident microflora and to any host plant.

Practically no considerations whatsoever has been given to the diversity of N_2 fixing bacteria available for inoculation. The choice of a bacterial species often appears as a prejudice without experimental or rational basis.

Our approach of inoculation is different : its first principle is to look for strains to inoculate among bacteria abundant in the rhizosphere of an actively fixing plant growing on the soil to be used. This warrants a good colonizing capability of the inoculum. The second principle is to choose the optimal plant-bacterium combination in preliminary laboratory studies.

In the case reported here we could in vitro select *Azospirillum lipoferum* 4B and mutant 17 of rice cv Césariot as a promising combination to use in Camargue. Subsequent field experiments confirmed that a positive effect was obtained when inoculating this particular rice genotype by strain 4B, except in conditions under which denitrification is likely to have occurred.

More research is currently in progress in the same direction using the same rationale to select rice and bacterial genotypes appropriate for inoculation of rice in different countries and climates and on different types of soils.

Acknowledgements

This research has been financially supported by the french Délégation Générale à la Recherche Scientifique et Technique and the private fertilizer company CdF-Chimie.

We are particularly indebted to Dr. A. MARIOTTI, Université Paris VI and Pr. R. BARDIN, Université Lyon I who analysed our samples for their isotopic composition.

References

1. Amarger, N., Mariotti, A. and Mariotti, F. (1977). C.R. Acad. Sci. Paris 284, 2179-2182.

2. Amarger, N., Mariotti, A., Mariotti, F., Durr, J.C., Bourguignon, C. and Lagacherie, B. (1979). Plant Soil 52, 269-281.

3. Balandreau, J. (1982). Mutation Breeding News Letter, I.A.E.A. 20, 4-5.

4. Bally, R., Thomas Bauzon, D., Heulin, T., Balandreau, J., Richard, C. and De Ley, J. (1983). Can. J. Microbiol. 29, 881-887.

5. Bardin, R., Domenach, A.M. and Chalamet, A. (1977). Rev. Ecol. Biol. Sol 14, 395-402.

6. Domenach, A.M., Chalamet, A. and Pachiaudi, C. (1979). C.R. Acad. Sci. Paris Série D 289, 291.

7. Dommergues, Y., Balandreau, J., Rinaudo, G. and Weinhard, P. (1973). Soil Biol. Biochem. 5, 83-89.

8. Holmes, B., Owen, R.J., Evans, A., Malnick, H. and Willcox, W.R. (1977). Int. J. Syst. Bacteriol. 27, 133-146.

9. Heulin, T., Bally, R. and Balandreau, J. (1982). Experientia Suppl. 42, 92-99.

10. Heulin, T., Weinhard, P. and Balandreau, J. (1983). Experientia Suppl. 48, 89-94.

11. Marie, R. (1970). In "Rice breeding with induced mutations II". I.A.E.A. Technical Reports Series 102, 21-24.

12. Marie, R. (1974). L'agronomie tropicale. 29, 892-900.

13. Pankhurst, E.S. (1967). Lab. Pract. 16, 58-59.

14. Rennie, R.J. (1979). Rev. Ecol. Biol. Sol 16, 455-463.

15. Thomas Bauzon, D., Weinhard, P., Villecourt, P. and Balandreau, J. (1982). Can. J. Microbiol. 28, 922-928.

16. Weaver, P.K., Wall, J.D. and Gest, H. (1975). Arch. Microbiol. 105, 207-216.

EVALUATION OF NITROGEN FIXATION IN SORGHUM CULTIVARS INOCULATED WITH DIFFERENT STRAINS OF AZOSPIRILLUM spp.

C. Christiansen-Weniger, R.M. Boddey and J. Döbereiner
Programa Nacional de Pesquisa em Biologia do Solo - EMBRAPA
Seropédica 23460, Rio de Janeiro, Brazil

Summary: Three cultivars of sorghum were grown in closed 7 l pots with the aerial part under free sunlight. Washed sand mixed with 1 % organic matter labelled with ^{15}N was used and the applied N-fertilizer was labelled with equal N^{15} % excess. Four strains of Azospirillum spp., isolated from sorghum roots were used as inoculants. The strains were resistant against certain antibiotics, so that their identification was possible upon reisolation. At different plant growth stages C_2H_2 reduction was assayed after sealing the pots around the stems and total N and the N^{15} dilution was measured at harvest. To study the establishment of the inoculated strains, bacterial numbers in the root system were counted by the MPN method. The results and their meaning for understanding the function of a bacterial root-association will be discussed. Keywords: Sorghum, cultivars, inoculations, azetylene reduction, ^{15}N-method.

Introduction

Until now the functioning of Gramineae-*Azospirillum* associations is not very well understood, but it has been reported in several papers that both the strain of the bacteria and the genotype of the plant have influence on the whole system. High variations between the acetylene reducing activity of sorghum cultivars were demonstrated, ranging from almost zero to more than 5 μmol C_2H_4 plant.h^{-1} (13). Also differences in the effectivity of associated *Azospirillum* strains were observed. One of the possible reasons for this

Azospirillum III: Genetics, Physiology, Ecology
Ed. by W. Klingmüller

was suggested to be the ability of some bacteria to infect and establish in the inner root tissue (2, 9).

In the present paper we report interactions of different plant cultivars with *Azospirillum* strains isolated from within roots or from the root surface in sorghum grown in pots with sand and ^{15}-N labelled organic matter.

Materials and Methods

A pot experiment with a randomized complete block design was planted, including 4 replications, 3 sorghum cultivars and 5 inoculation treatments. The cultivars used were CMSXS 136 and CMSXS 114, both tolerant to aluminium toxicity, and BR 007b which is know to be aluminium sensitive. The parameter of aluminium tolerance was chosen because one aluminium tolerant cultivar was shown to contain a higher organic acid content in the root (mainly malic and trans-aconitic) than that of an aluminium sensitive cultivar (5). The plants were grown in plastic pots containing 7 kg of washed sand mixed with 1 % organic matter (dry, finely-ground, ^{15}-N labelled grass with 0.83 % N). Nutrients were added to the pots as follows: 53 ppm P, 66 ppm K and 1 ml kg^{-1} of trace element solution (8). Mineral nitrogen was added every 2 days (20 mg N per addition) as ^{15}-N labelled KNO_3. Both the organic matter and the potassium nitrate had the same ^{15}-N label of 0.2855 atom % excess. The experiment stood outside until panicle emergence and was then transferred to a temperature controlled greenhouse. For acetylene reduction assays the pots were closed with lids and 10 % v/v acetylene was injected through a septum in the container bottom as described by Wani *et al.* (13). Gas samples were taken after 2.5 and 6 hours and analysed for C_2H_4 by gas chromatography. To eliminate residual acetylene, which initially was found to persist in the sealed pots for several days, after the 3rd acetylene reduction assay the pots were blown through with air after assay. The plants were at an early stage of maturity and divided into root, stem plus leaves, and panicle for subsequent drying and analysis for total nitrogen and ^{15}-N enrichment (3).

For inoculation four strains of *Azospirillum* were used:

Strain	Origin
S 82 *A. lipoferum*	Surface sterilized sorghum roots (1% chloramine T, 30 min)
S 19 *A. lipoferum*	Surface sterilized sorghum roots (1% chloramine T, 30 min)
S 65 *A. lipoferum*	Washed sorghum roots
Sp 7 *A. brasilense*	ATCC 29145 from rhizosphere soil of *Digitaria decumbens*

The strains were grown 24 hours at 30°C in liquid NFb medium (1) supplemented with 0.5 g.L^{-1} NH_4Cl. One ml of this suspension was added directly to 7 day old seedlings and control pots were inoculated in the same way with an autoclaved bacterial culture.

Immediately after harvest MPN counts (McCrady) of the bacterial root population were made using semi-solid NFb medium containing 5 g malate and 2.5 g glucose L^{-1} (6).

Results and Discussion

After the stage of panicle emergence, acetylene reduction activity was measured every week. As demonstrated in Fig. 1 the activity over time showed similar patterns for all three cultivars (for figure 1 see end of article).

It seems that the main activity develops only in the late phase of plant growth especially after flowering. This agrees with other work on cereal associations (4, 12, 14), although the reason for it is not yet known. Possibly during flowering and grain filling production and translocation of organic acids in the plant increases so that at this time sufficient carbon sources are available for bacterial metabolism. The depression of nitrogenase activity during grain filling (arrow) could be explained by the competition for photosynthates between the panicle and the root-associated bacteria.

The cultivar BR 007b, which is sensitive to aluminium toxicity showed over all assays a significantly lower acetylene reducing activity than the aluminium tolerant cultivars. This seems to confirm expected differences in the organic acid content in the root systems (5).

The effect of bacterial inoculation on the acetylene reduction activity is shown in Table 1. Significant differences in activity were observed between plants inoculated with strain S 82 isolated from inside the roots and S 65 isolated from the root surface or Sp 7 isolated from rhizosphere soil.

That the inoculated control treatment also developed high acetylene reducing activity must be attributed to the fact that the growth medium was not sterilized and contamination with native *Azospirillum* occurred. This is confirmed by the MPN

Table 1. Effect of the inoculation treatments on acetylene reduction activity. N content and ^{15}-N dilution in the plant. The values for ARA are means of 5 assays. Means of the three cultivars

	ARA µmol C_2H_4 $h^{-1}pot^{-1}$	Total N mg $plant^{-1}$	Top		
			N content %	N^{15} % excess	% N from fixation
S 82	3061 a	518 a	1,17 a	0,2797 a	2,03 b
S 19	2245 abc	499 ab	1,10 ab	0,2799 a	1,96 b
S 65	1397 c	467 b	1,07 ab	0,2788 ab	2,35 ab
SP 7	1967 bc	467 b	1,02 b	0,2777 b	2,73 a
Control	2554 ab	465 b	1,06 b	0,2784 ab	2,49 ab
F value	**		*	*	*

** P = 0,01, * P = 0,05, n.s. and values with the same letter are not significantly different.

counts, which showed high numbers of *Azospirillum* in the roots (mean 2.5 x 10^6 cells per g fresh root) of all treatments including the controls without significant differences between cultivars or inoculation treatments.

Total N content and % N data in the plant are shown in Tables 1 and 2. Unfortunately the three sorghum cultivars were of different phenotypic growth habits and therefore there is a large difference between their dry matter production and the periods in which they are actively taking up nutrients. Because of this, comparison between cultivars to assess nitrogen fixing activity from differences in N accumulation is difficult.

Table 2. Dry matter production, N content and ^{15}-N dilution of the three sorghum cultivars. Means of all inoculation treatments

	Top			Root	Top	
	dry matter g	N total mg	N content %	N content %	N^{15} % excess	% N from fixation
CMSXS 136	39,9 b	487,2	1,22 a	0,40 b	0,2788 b	2,35 a
CMSXS 114	48,0 a	471,3	0,98 c	0,51 a	0,2781 b	2,59 a
BR 007b	45,6 b	498,7	1,09 b	0,36 b	0,2800 a	1,96 b
F value test	**	n.s.	**	**	**	**

** P = 0,01, * P = 0,05, n.s. and values with the same letter are not significantly different.

A comparison of the inoculation treatments shows that the strains S 82 and S 19, both isolated from within roots of sorghum, caused the highest values of total N and N percent in the plant. Similar observations were made in a wheat experiment where the total N accumulation in the plant tops correlated significantly with the numbers of *Azospirillum* in chloramine T treated roots, but not with the *Azospirillum* numbers in the unsterilized roots (2).

To quantify the contribution of nitrogen fixation to the plant N assimilation the ^{15}-N enrichment was determined. As all the available nitrogen in the growth medium had the same ^{15}-N enrichment, any dilution of the ^{15}-N label in the plant should be caused by an input from atmospheric nitrogen. These results are reported in Tables 1 and 2. Amongst the cultivars, BR 007b showed the highest value for % ^{15}-N content and therefore the lowest contribution of nitrogen from nitrogen fixation. This difference is significant and confirms the estimates made with the acetylene reduction method (Fig. 1). The relatively low proportion of fixed nitrogen in the plant tops (2 to 2.6% of total N) may be caused by the fact that most nitrogen fixing activity in the roots starts only in a late phase of plant growth, when most nitrogen has already been taken up from the growth medium.

The effects of the inoculation treatments on ^{15}-N dilution are more difficult to explain than those of the cultivars. Strains S 82 and S 19, which caused the highest nitrogen content in the plant, had the lowest value for ^{15}-N dilution. On the other hand, strain Sp 7, although it has only a small effect on plant nitrogen shows in all inoculation treatments the highest contribution of nitrogen derived from nitrogen fixation. The explanation for these apparently contradictory results must be that the associated bacteria, besides their ability to fix nitrogen, interact with the uptake of mineral nitrogen from the soil. This could be a) an indirect effect, where bacterially produced plant growth substances caused an increase in the root system as described by Tien *et al.* and Okon (10,11) or b) a direct participation of the associated bacteria in the nitrate-reducing pathway of the plant as suggested by Döbereiner (7).

The mean ^{15}-N dilution values of roots, stem plus leaves, and panicle are shown separately in Table 3. The ^{15}-N enrichment in the roots indicates that approximately 26 % of the nitrogen in the roots was derived from associated biological nitrogen

Table 3. Mean ^{15}-N dilution and percent nitrogen from fixation in panicle, stem plus leaves, and root. The label of the mineral nitrogen source was 0.2855 atom % excess

	% N^{15} excess	% N from fixation
Panicle	0,2845	0,35
Stem, leaves	0,2735	4,20
Root	0,2115	25,90

fixation, as opposed to 4 % in the leaves and stem and virtually no contribution to the panicle. It seems probable therefore that the nitrogenase activity is used initially only for bacterial growth in the rhizosphere and very little fixed N is translocated to the shoot immediately. Only in the late phase of plant growth is nitrogem from dead bacterial tissue translocated in some form to the leaves and stems of the plant.

Conclusion

Between sorghum and *Azospirillum* spp. exists and N-2 fixing association of relatively high acetylene reduction activity. The acetylene reduction data indicate that the cultivars resistant to aluminium toxicity had higher associated N-2 fixing activity than a aluminium sensitive cultivar which may be related to the higher organic acid content of the roots of such cultivars (5). Some nitrogen from fixation was translocated to the shoot of the plant.

Between the different inoculation treatments the strains isolated from within roots had a significantly greater effect on plant nitrogen than strains isolated from washed roots or rhizosphere soil. It was demonstrated that the associated *Azospirillum* besides having the ability to fix nitrogen also influenced the assimilation of nitrate by the plant.

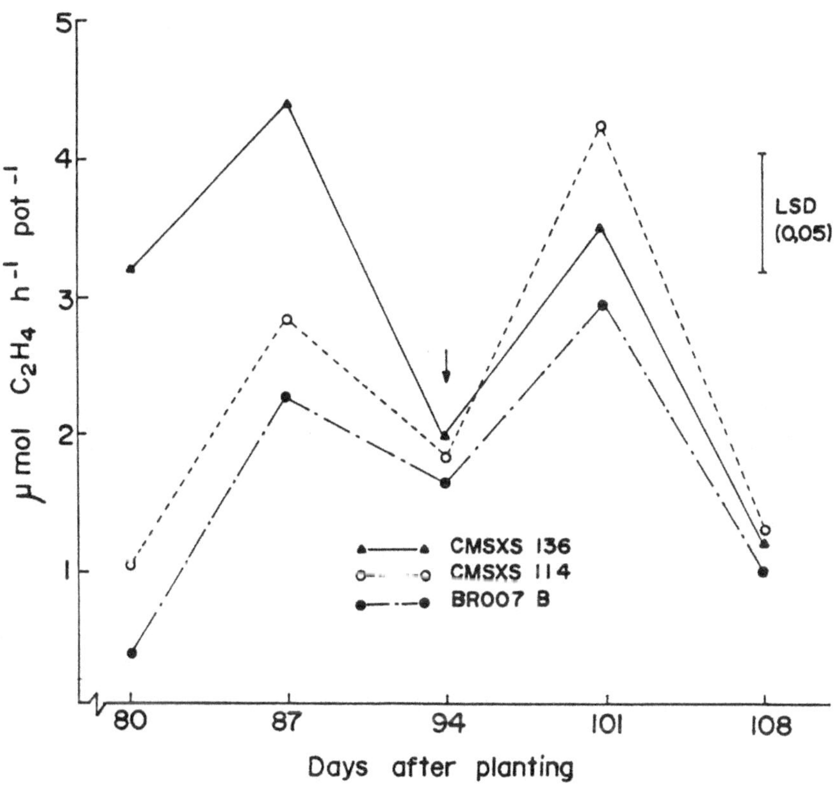

Fig. 1. Acetylene reduction activity of the three sorghum cultivars measured over 5 weeks. Vertical arrow indicates mid grain filling stage.

References

1. Baldani, V.L.D. and Döbereiner, J. 1980. Soil Biol. Biochem. 12, 433-439.
2. Baldani, V.L.D., Baldani, J.I. and Döbereiner, J. 1983. Can. J. Microbiol. 29, 924-929.
3. Boddey, R.M., Chalk, P.M., Victoria, R. and Matsui, E. 1983. Soil Biol. Biochem. 15, 25-32.
4. Bülow, J.F.W. von and Döbereiner, J. 1975. Proc. Nat. Acad. Sci. 72, 2389-2393.
5. Cambraia, J., Galvani, F.R., Estevão, M.M. and Sant'anna, R. J. Pl. Nutr. 6, 313-322.
6. Döbereiner, J. 1980. In: Methods for Evaluating Biological Nitrogen Fixation (F.J. Bergersen, ed.) John Wiley & Sons Ltd., 535-555.
7. Döbereiner, J. 1983. In: *Azospirillum* II (W. Klingmüller, ed.) Experientia Supplementum, 48, Birkhäuser Verlag Basel, 9-23.
8. Franco, A.A., Peres, J.R.R. and Nery, M. 1978. Plant and Soil. 50, 1-11.
9. Magalhães, F.M.M., Patriquin, D. and Döbereiner, J. 1979. Rev. Brasil. Biol. 39, 587-596.
10. Okon, Y. 1982. Israel J. Bot. 31, 214-220.
11. Tien, T.M., Gaskins, M.H. and Hubbell, D.H. 1979. Apl. Environ. Microbiol. 37, 1016-1024.
12. Wani, S.P., Dart, P.J. and Upadhyaya, M.N. 1983. Can. J. Microbiol. 29, 1063-1069.
13. Wani, S.P., Upadhyaya, M.N. and Dart, P.J. 1984. Plant and Soil, 82, 15-29.
14. Watanabe, I., Barraquio, W.L., Guzman, M. R. de and Cabrera, D.A. 1979. Apl. Environ. Microbiol. 37, 813-815.

POSSIBLE CONTRIBUTION OF AZOSPIRILLUM spp. TO THE NUTRITIONAL STATUS OF WHEAT PLANTS GROWN IN SANDY SOILS OF GASSIM-SAUDI ARABIA

N.A. Hegazi+ and H. Saleh
College of Agric., King Saud Univ., Gassim, Buraydah
P.O. Box 1482, Saudi Arabia

Summary: Representative strains of *Azospirillum* spp. were isolated from roots of plants grown in Gassim, Saudi Arabia. They were identified as *A. brasilense*. A liquid inoculum prepared from a mixture of 3 potent strains was used to inoculate seeds of 2 wheat cultivars. A field trial was executed to compare biofertilization with azospirilla against N-fertilization. Significant responses to inoculation in the presence of a limited supply of N was obtained particularly for cv. West-Breed. The root biomass was very significantly increased and there were substantial increases in straw yield. Seed yield was 31.5 % higher than in uninoculated plants receiving a limited dose of N.
Keywords: Biofertilization, inoculation, associative N_2-fixation, wheat-association, diazotrophs.

Introduction

Soils of the Gassim area, in the centre of Saudi Arabia, are virgin sandy soils belonging to Aridisol. An active agricultural extension programme is underway in the Kingdom and substantial areas of such soils are being brought under cultivation with wheat as a major crop. Soil analyses, including microbiological analysis, are needed to evaluate fertility of the soils under reclamation as well as to help in selection of appropriate crops and in applying suitable fertilization regimes. Hegazi and Al-

---

+Author to whom all correspondence should be made. (Permanent address: Fac. Agric., Cairo Univ., Giza, Egypt).

Azospirillum III: Genetics, Physiology, Ecology
Ed. by W. Klingmüller

Sahael (6) presented data on microbiological analysis of soils of various localities. In general, soils contained limited numbers of diazotrophs (10^3 g^{-1}). Certain groups of such bacteria, particularly clostridia and azospirilla, are found to be enriched to a certain extent on roots of various plants grown in the area. Representative isolates of *Azospirillum* spp., in particular, were isolated and studied for their different characteristics. They were employed in the present study to inoculate seeds of wheat, with or without N-fertilization. This is an effort to see how far *Azospirillum* spp. might contribute to the growth and productivity of wheat plants cultivated under the extremely arid conditions of Saudi Arabia.

Materials and methods

Soils:

The present inoculation trial was conducted in the fields of the Experimental Station of the College of Agriculture, King Saud University, at Buraydah, Gassim. The soil is loamy-sand and it has the following composition: Organic carbon (%), 0.02-0.07; $CaCO_3$ (%), 5.5-8.7; available $PO^{=}_4$ (ppm), 14-18; toal-N, Nil; T.S.S. (ppm), 560; pH, 8.2. Microbiological analysis indicated that the soil possessed the following population densities (per g soil): Total bacteria, 4.8 x 10^5; total spore-forming bacteria, 3.4 x 10^4; ammonifiers: 2 x 10^5; actinomycetes, 10 x 10^2, cellulose decomposers, 4.5 x 10^2; denitrifers, 25 x 10^2 and nitrifiers, 1.6. x 10^2.

Seeds:

Local (untreated with fungicides) and imported seeds (pre-treated with the fungicide Carboxin) of two American cultivars of wheat (*Triticum aestivum*), Yoco-Rogo and West Breed, were obtained. The viability of cells of azospirilla present on inoculated seeds - with or without fungicide treatment - were compared during 7 days-storage at room temperature (20-35° C) for cv. Yoco-Rogo.

Inocula:

Three strains of Azospirillum spp. - identified as A. brasilense - were isolated from roots of Sudan grass. They were grown separately in semi-solid N-deficient malate medium for 72 hrs. A mixture of bacterial suspensions (1.5×10^8 cells ml^{-1}) was prepared and used for coating the seeds, which were left to dry over night. Numbers of azospirilla on seeds were estimated by the m.p.n. technique, and those seeds used for sowing had 35×10^4 cells $seed^{-1}$.

Layout of the field experiments:

A split-plot design in a randomized complete block with three replications was used. Wheat cultivars were assigned to the main plots. There were four treatments: 1) azospirilla inoculation; 2) full dose of N only (200 kg N ha^{-1}); 1/3 dose of N only (67 kg N ha^{-1}); 4) simultaneous azospirilla inoculation + 1/3 dose of N. Treatments were distributed randomly in sub-plots. The field was fertilized with 75 kg phosphorous (calcium triple phosphate 45-47 % P_2O_5) per hectare during seed bed preparation. N-fertilization, in the form of urea 45 % N, was applied in 2 doses, at planting and after 8 weeks. The plot size was 7.2 m^2 consisting of 11 rows 4 m long and 15 cm apart. On 3.1.1985. seeds were sown and then sprayed, before covering, with additional bacterial suspension (2×10^6 cells ml^{-1}) either alive (for inoculated treatments) or autoclaved (for all others). General agronomic practices were used. Periodic samples were taken from the outer four rows while the inner seven ones were kept for harvest. Total dry weight (at 70° C) of plants, roots and ears, as well as numbers of azospirilla, in free soil and on roots, were recorded for 2 month-old plants. At harvest, seed and straw yield were obtained for each plot. Total nitrogen content was determined by Kjeldahl analysis (2).

Microbiological analysis:

Ten-fold dilutions of soil samples were prepared in sterile distilled water. Bacteria associated with plant roots were counted initially in first dilutions prepared by crushing roots with closely-adhering soil in sterile distilled water using a Waring blender. Enumeration of Azospirillum, as well as other groups of diazotrophs, was carried out using the technique of most probable number (m.p.n.) and the respective semi-solid media (4,7,10).

Statistical analysis:

Results were analysed statistically using the analysis of variance.

Results

Soils (Aridisol) under investigation were coarse-textured with a pH in the range of 7.4 - 8.5, and contained an average microbial population of 30 - 559 x 10^5 g^{-1} soil. The majority (79 %) are spore-forming bacteria. Dinitrogen fixers are present in limited numbers (0.5 - 90 x 10^2 g^{-1}). *Azospirillum* spp., *Clostridium* spp., and N_2-fixing bacilli and other Gram-negative short rods were encountered but *Azotobactr* was not. These particular groups of microorganisms were found also on roots of various plants grown in the area, e.g. wheat and alfalfa. Representative isolates of *Azospirillum* spp. were isolated from roots of Sudan grass and were identified as *A. brasilense*. They were further used for the present inoculation field trial.

Fig. 1 presents the results of a preliminary experiment conducted to test for the viability of azospirilla cells present on seeds with (imported) or without (local) treatments of the fungicide Carboxin. Numbers of viable cells on seeds, after coating with the liquid inocula of azospirilla, decreased with time. Such decreases were much more pronounced (66.7 %) with treated than with non-treated seeds. Therefore, it was decided to use the local seeds for the inoculation trial in which N-fertilization and azospirilla inoculation were compared for two cultivars of wheat.

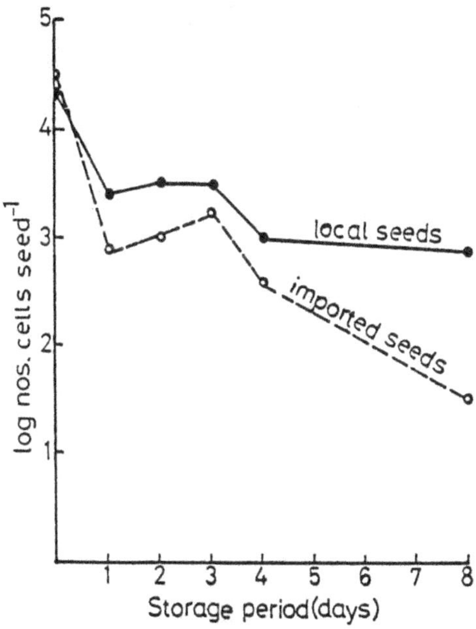

Fig. 1. Survival of cells of azospirilla on seeds treated (imported) and non-treated (local) with the fungicide Carboxin.

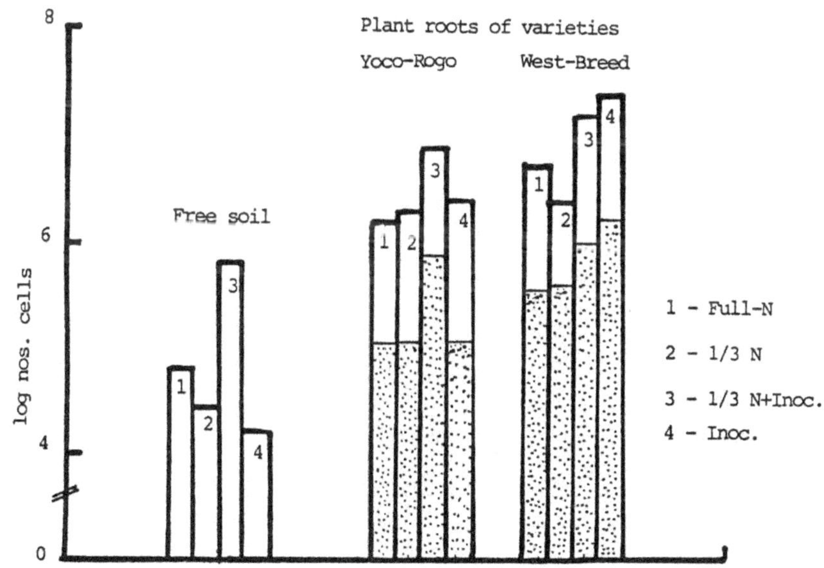

Fig. 2. Densities of azospirilla on plant roots (▭ per g, and ▒ per plant) as compared to those found in free soil.

The development of plants of various cultivars was followed during vegetative growth. Samples of 8 week-old plants were obtained and analysed for a number of parameters. Counts of azospirilla were estimated (Fig. 2) and bacteria were found on roots with higher numbers (10^6 cells g^{-1} dry matter) than in free soil. Irrespective of treatment, roots of cv. West-Breed accommodated higher densities of azospirilla than cv. Yoco-Rogo. Inoculation increased numbers of bacteria on the plant roots particularly those of c.v. West-Breed. Weight of plants of this particular age differed significantly among treatments (Fig. 3). N-fertilized plants as well as those given both azospirilla and a limited dose of N were superior to plants of other treatments. Differences among cultivars were not statistically significant. Weight of plant roots correlated significantly with total dry weight of plant (Fig. 4a). The differences in root biomass attributed to either cultivars or to various treatments were highly significant (Fig. 4b). The cultivar West-Breed possessed a larger root biomass than cv. Yoco-Rogo, particularly when a

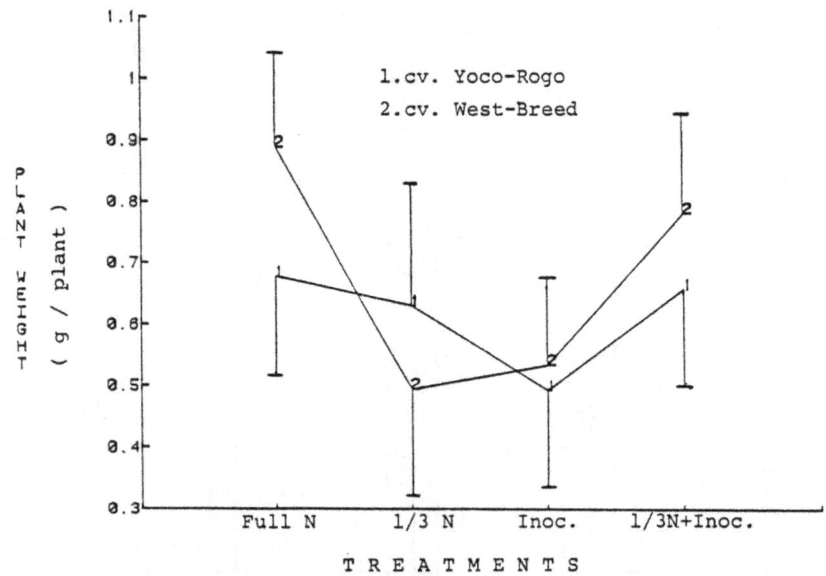

Fig. 3. Dry weight of wheat plants as a affected by various N-fertilization and inoculation treatments.

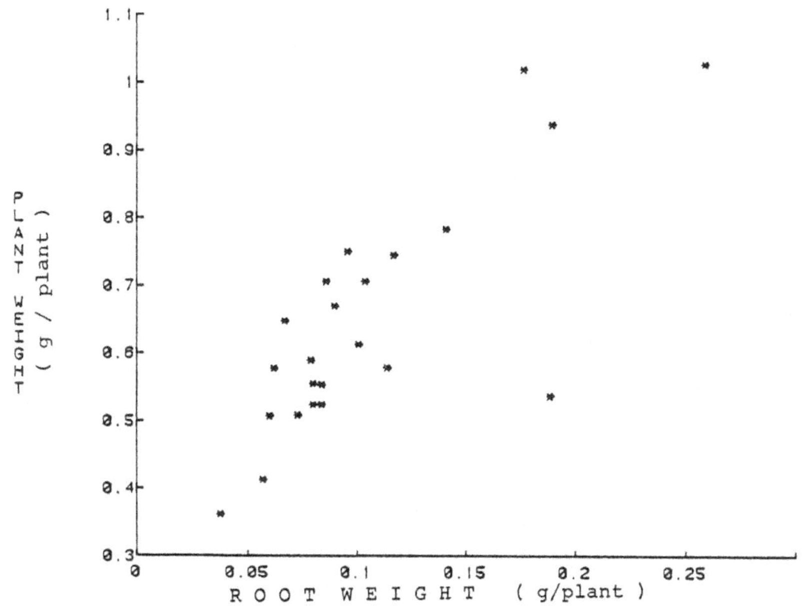

Fig. 4a. Scatter diagram of plant weight and root weight of wheat plants irrespect of varieties and treatments.

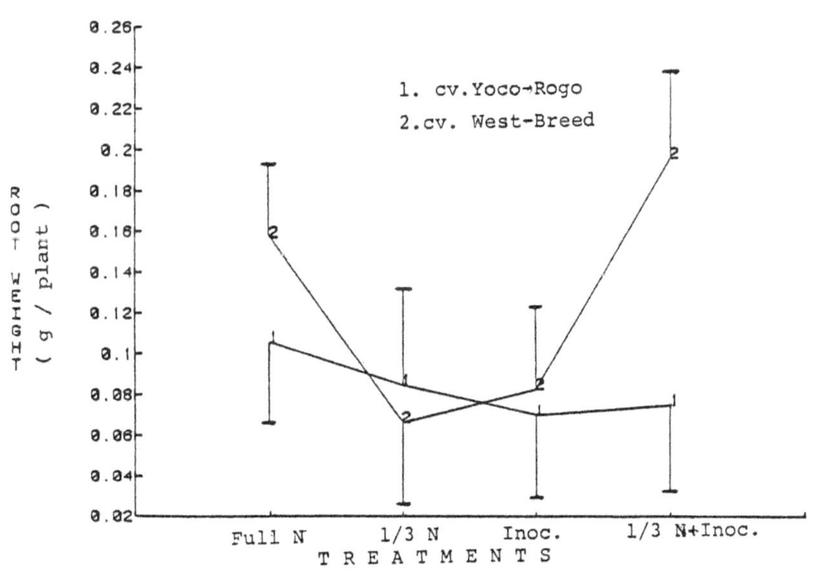

Fig. 4b. Effect of various treatments on weight of roots of wheat plants.

limited dose of N was given simultaneously with azospirilla inoculation. Weight of ears differed among treatments but not among varieties (Fig. 5), and was highest for N-fertilized plants and for those 1/3 N-fertilized together with azospirilla inoculation.

Yields of wheat cultivars were obtained and are illustrated in Figs. 6 and 7. In general, both seed and straw yields correlated very significantly (Fig. 6a). The differences in straw yield recorded among cultivars as well as among treatments were significant. The cultivar West-Breed gave higher yields than cv. Yoco-Rogo, and both full N-fertilization and simultaneous inoculation and application of a limited dose of N gave similar yields, higher than those obtained with other treatments. Results obtained for seed yield (Fig. 7) indicated that differences among treatments were significant and those between varieties approached significant levels ($p = 0.094$). Highest yields were recorded for cv. West-Breed with full N-fertilization or

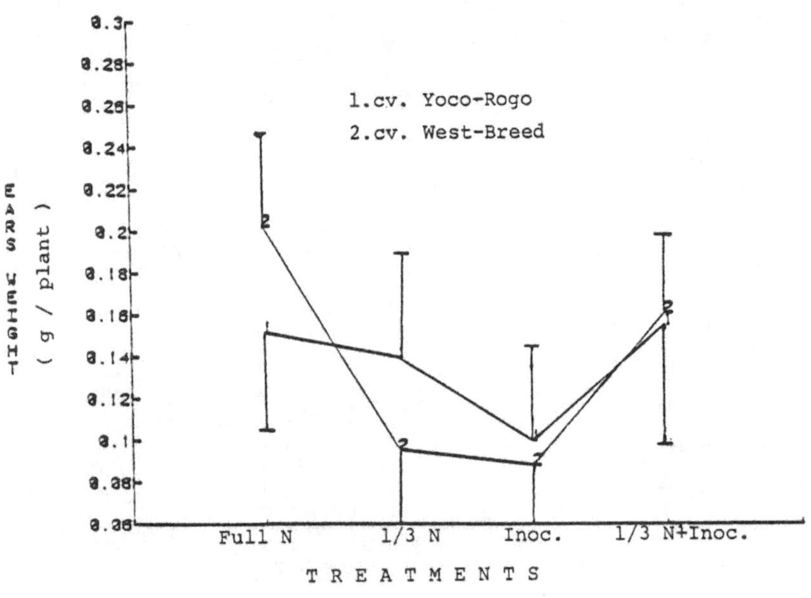

Fig. 5. Effect of various N-fertilization and inoculation treatments on the weight of ears of wheat plants.

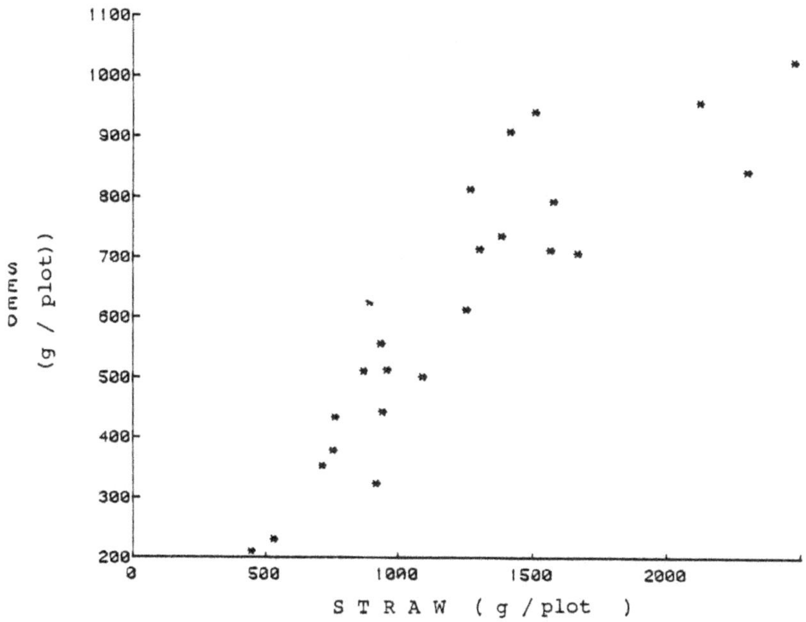

Fig. 6a. Scatter plot of both seed and straw yields.

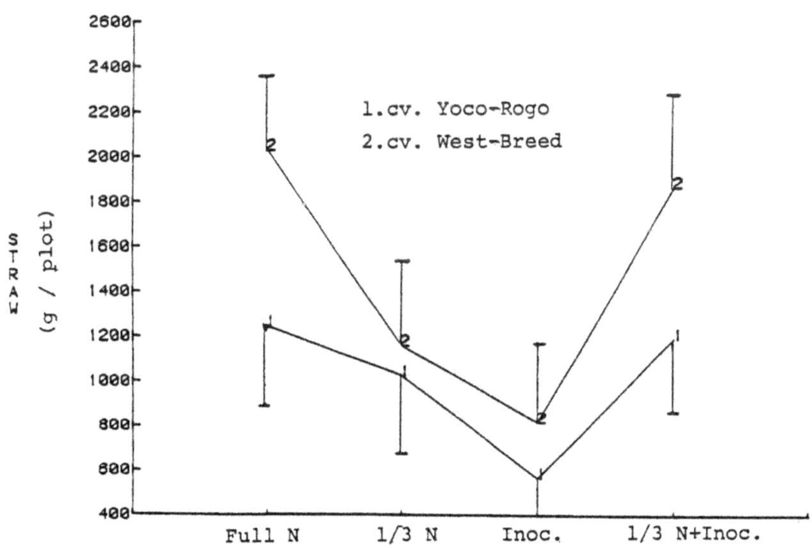

Fig. 6b. Straw yield as affected by inoculation and/or N-fertilization treatments.

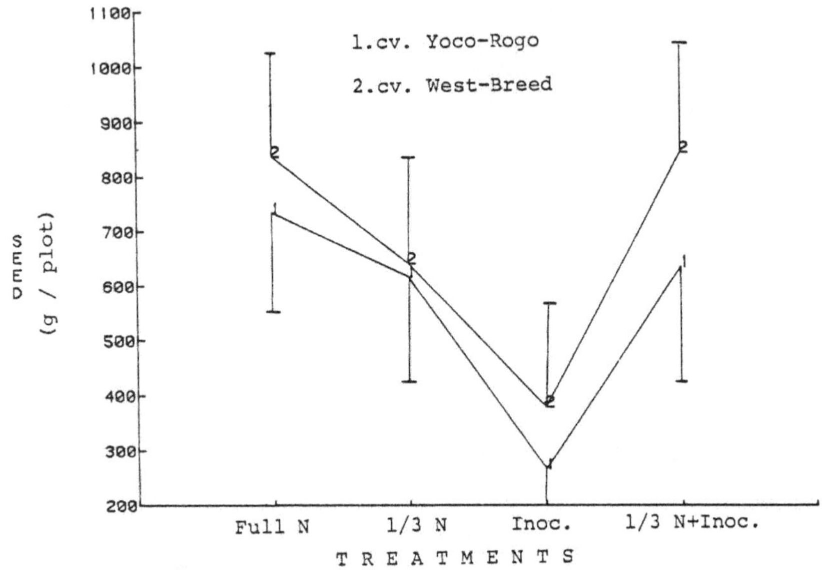

Fig. 7. Response of seed yield to various N-fertilization and inoculation treatments.

azospirilla inoculation together with application of a limited dose of N. Estimates of total N-content of plants (0.4 - 0.77 %) of either cultivar and of various treatments did not vary significantly.

Discussion

Among the beneficial effects of rhizosphere microflora is their contribution to the nutritional status of growing plants. This is particularly important under extreme conditions such as those prevailing in poor sandy desert soils. Various groups of microorganisms including diazotrophs were reported enriched on the roots of a number of plants grown in such soils, to the extent that it is quite possible that they contribute to the N-status of plants (9,17). Bringing sandy soils, including those under investigation, under cultivation requires an intensive regime of fertilization. However, manipulation of diazotrophic microorganisms for the benefit of either leguminous or non-leguminous plants offers an additional and sometimes economical source of nutrients. In the last few years there have been reports of substantial and economically significant increases in yield (5 - 30 %) and total N of field grown cereal crops (wheat, maize, sorghum and pearl millet) following inoculation with *Azospirillum* spp. in particular (1,3,4,5,8,10,15,20). Results of the present field trials confirm the significance of wheat inoculation with *Azospirillum* in soil where no or few azospirilla cells occur and where wheat has not been planted before. Increased development of inoculated plants and yield increases were reported particularly in the presence of a supporting limited dose of N (ca. 67 kg n ha^{-1}). The presence of a low dose of N is among the factors which help in establishing associative N_2-fixation on plant roots and which boost the effect of inoculation as reported for sorghum (4) and other cereal crops (13). Among other factors that might have contributed to the accummulative positive response of azospirilla inoculation reported in this study is the use of specifically selected potent strains of *Azospirillum*. This is a point that has been emphasized by Baldani *et al*. (1). Additionally, the inoculum used was in liquid form and ranged from 10^4 - 10^5 cells pers plant, which has been reported to be appropriate for inoculation of grasses (4,8,10,13).

The present inoculation trial has identified a potentially beneficial host-*Azospirillum* combination. The cultivar West-Breed responded positively to azospirilla inoculation while cv. Yoco-Rogo did not (Table 1). Increases attributed to inoculation in the presence of a limited dose of N were significant for both dry matter and straw yield. The grain yield was 31.5 % higher than in the uninoculated control, but this difference was not significant. Of particular interest was the very significant increase in root biomass attributed to azospirilla inoculation. Other investigators have obtained similar results with wheat and other cereal crops (1,4,10,13). Studies by Patriquin *et al*. (16)

Table 1. Response of wheat plants (West-Breed[a]) to azospirilla inoculation in presence of a limited dose of N-fertilization

Parameter	N-fertilization (67 kg N/ha)	Inoculation+ 67 kg N/ha	Levels of significance
Nos.azospirilla ($x\ 10^3$/plant)	526	933	Nd.[b]
Plant weight (g)	0.495	0.784	0.058[c]
Root weight (g)	0.066	1.196	0.005**
Ears weight (g)	0.095	0.161	0.122
Seed yield (g/plot)	642	844	0.154
Straw yield (g/plot)	1159	1875	0.030*

a, no significant responses were obtained with cv. Yoco-Rogo; b, not analysed; c, approached significant levels.

on root hair deformation associated with azospirilla inoculation illustrated pronounced effects of the plant genome and bacterial genome, at the cultivar and strain levels respectively. Further studies (11,12,20) indicated much more root hair branching and deformation in the presence of Azospirillum than in the presence of non-Azospirillum strains. This effect was a strain-specific one and differed among different cultivars. The increase of root biomass - in the case of cv. West-Breed in particular - for azospirilla-inoculated plants is thought to be associated with the production of hormones by Azospirillum (15,19). Recently, Jain and Patriquin (12) were able to isolate and identify IAA from culture filtrates of Azospirillum which caused branching of root hairs similar to that caused by the Azospirillum itself. It has also been demonstrated that Azospirillum which live on or in the root can produce phytohormones affecting plant growth by inducing the proliferation of lateral roots. The very significant increase in root biomass reported in this study, as well as in others, particularly suggest the role of growth hormones. In addition, the increase in root surface may have resulted in more efficient nutrient uptake (15). However, one cannot exclude absolutely the ability of azospirilla to fix N_2 and consequently enhance N assimilation by plants (1,4,10). Positive response to inoculation of wheat with diazotrophs other than Azospirillum has also been recorded and involvement in N_2-fixation by Bacillus spp. was confirmed by ^{15}N-isotope dilution (18).

Acknowledgements

The cooperation of Prof. W. Klingmüller of the Department of Genetics, University of Bayreuth, and the financial support of the Alexander von Humboldt Foundation, Bonn, West Germany, was a kind gesture which made the participation of the first author in the Workshop possible. Statistical analysis was carried out with the collaboration of R. Brandl of the Department of Animal Ecology, University of Bayreuth. We thank A. Hussein, O. Abou-Geab and M. El-Dochael for technical assistance. We appreciat the assistance of the Soil Science students, S. Al-Sahw, S. Al-Meshari and H. Al-Atiq, of College of Agriculture, King Saud University at Gassim.

References

1. Baldani, V., Baldani, D., Ivo Baldani, J. and Döbereiner, J. (1983). Can. J. Microbiol. 29: 924-929.
2. Bremner, J.M. (1965). In Methods of soil analysis. II. Chemical and microbiological properties. American Society of Agronomy, Madison, WI., pp. 1149-1178.
3. Cohen, W., Okon, Y., Kigel, J., Nur, I. and Henis, Y. (1980). Plant Physiol. 66: 746-749.
4. Eid, M.A. Hegazi, N.A., Monib, M. and Shokr, E. (1984). Rev. Ecol. Biol. Sol. 21: 235-242.
5. Hegazi, N.A. (1983). In Azospirillum II, Genetics, Physiology, Ecology. EXS 48, pp. 171-189.

6. Hegazi, N.A., Al-Sahael, Y. (1983). Egyptian Society of Applied Microbiology. Proceedings of V. Conf. Microbiol., Cairo, May 1983, pp. 49-63.
7. Hegazi, N.A., Amer, H. and Monib, M. (1979). Soil Biol. Biochem. 11: 437-438.
8. Hegazi, N.A., Khawas, H. and Monib, M. (1981). Proceedings of IV. Int. Sym. on Nitrogen Fixation. Canberra, Australia. Australian Academy of Sciences Publications, pp. 493.
9. Hegazi, N.A., El-Mallawani, A.A. and Monib, M. 1980. Egyptian Society of Applied Microbiology. Proceedings of IV. Conf. Microbiol., Cairo (1980), pp. 119-123.
10. Hegazi, N.A., Monib, M., Amer, H.A. and Shokr, E. (1983). Can. J. Microbiol. 29: 888-894.
11. Jain, D.K. and Patriquin, D.G. (1984). Appl. Environ. Microbiol. 48: 1208-1213.
12. Jain, D.K. and Patriquin, D.G. (1985). Can. J. Microbiol. 31: 206-210.
13. Kapulnik, Y., Sarig, S., Nur, I., Okon, Y., Kigel, J. and Henis, Y.H. (1981). Expl. Agric. 17: 179-187.
14. Kapulnik, Y., Sarig, S., Nur, I. and Okon, Y. (1983). Can. J. Microbiol. 29: 895-899.
15. Okon, Y. (1984). In Advances in nitrogen fixation research. Proceedings V Int. Sym. on Nitrogen Fixation. Noordwijkerhout, The Netherlands. Martinus, Nijhoff/Dr. W. Junk Publishers, pp. 303-309.
16. Patriquin, D.G., Döbereiner, J. and Jain, D.K. (1983). Can. J. Microbiol. 29: 900-915.
17. Rao, A.V. and Venjateswarlu, B. (1982). Can. J. Microbiol. 28: 778-782.
18. Rennie, R.J. and Larson, R.I. (1981). In Associative N_2-fixation. Edited by P.B. Vose and A. P. Ruschel. CRC Press, Palm Beach, Fl. pp. 145-154.
19. Tien, T.M., Diem, H.G., Gaskins, M.H. and Hubbel, D.H. (1981). Appl. Environ. Microbiol. 37: 1016-1024.
20. Venkateswarlu, B. and Rao, A.V. (1983). Plant and Soil 74: 379-386.

ROOT COLONIZATION AND PLANT GROWTH OF GRASSES AND CEREALS IN UNSTERILE SOIL AND NITROGENASE (C_2H_2) ACTIVITY OF STERILE CEREAL SEEDLINGS AFTER INOCULATION WITH AZOSPIRILLUM spp.

G. Jagnow
Institut für Bodenbiologie, Bundesforschungsanstalt für Landwirtschaft, Bundesallee 50, 3300 Braunschweig, FRG

Summary: Increases of *Azospirillum* rhizosphere populations and of shoot growth were observed after inoculation of *Lolium perenne* and *Hordeum vulgare* seedlings with *Azospirillum lipoferum* and 3 weeks growth in pots in unsterile soil. After 7 weeks, increased shoot growth was observed with *Arrhenatherum elatius* and *Dactylis glomerata*, but not with *Lolium perenne* and *Festuca rubra*, *Azospirillum* populations being similar to uninoculated plants. With 4 wheat cultivars proportions of nitrogenase-positive sterile seedlings grown in semisolid agar after inoculation with *Azpspirillum* spp. (28°C) varied between 4 and 88%. Nitrogenase activity followed a log-normal distribution within rank oder of plants and was higher with *A. brasilense* than with *A. lipoferum*. With maize seddlings a linear relationship was shown between log N_2-ase activity and log shoot length increments. The model appears to be suitable to screen for optimal "nis" phenotypes within cultivars.
Keywords: Azospirillum, rhizosphere, associative nitrogen fixation, screening, plant cultivars, plant growth.

Introduction

Yield increases of grasses and cereals were observed most frequently in tropical and subtropical climates after seed or seedling inoculation with *Azospirllum brasilense* in pot (1,2,3) and field experiments (1,4,5,6,7,8,9). In temperate climate, *A. lipoferum* was not yet tested for induction of plant growth responses, although a positive yield response was reported after its inoculation to wheat from a field experiment in India (10). Associative N_2-fixation appears to depend on the expression of several

Azospirillum III: Genetics, Physiology, Ecology
Ed. by W. Klingmüller

plant genes resulting in nitrogen fixation supporting ("nis") phenotypes (3,11,12,13). We therefore tested the effect of A. lipoferum in pot experiments and developed a screening method for nitrogenase activity with sterile grown cereal seedlings after inoculation with Azospirillum spp. to see whether varietal difference of the "nis" character also exist between German cereal cultivars.

Materials and methods

Plant cultivars: Seeds of Lolium perenne cv. NFG, Festuca rubra cv. Roland 21, Arrhenatherum elatius cv. Arel 41, Dactylis glomerata cv. Baranta, Zea mays cv. Anjou and cv. Mutin were obtained from the Institut für Grünland- und Futterpflanzenforschung, FAL, seeds of Hordeum vulgare cv. Oriol (summer barley), cv. Diana (winter barley), Triticum aestivum cv. Max, cv. Kadett, cv. Koral (summer wheat), cv. Disponent (winter wheat), Secale cereale cv. Teku (summer rye) and cv. Kustro (winter reye) were obtained from the Institut für Pflanzenbau und Pflanzenzüchtung, FAL.

Pot experiments: Unsterile barley and maize seeds were germinated in the dark for 2 days, grass seeds for 3-5 days at room temperature on moist filter paper. Healthy seedlings with 15-30 mm roots were selected and inoculated by 10 min. immersion into a cell suspension of Azospirillum lipoferum (14), obtained by washing growth from nutrient agar slants incubated for 2 days at 25°C with sterile deionized water ($O.D._{540}$ ca. 1.6). Control seedlings were immersed in sterile water. Plastic pots (13 cm diameter) with 4 drainage holes covered with nylon gauze were filled with a 50% (v/v) mixture of a parabrown earth (15) and fine quartz sand, soaked in tap water and drained. Two maize, four barley and ten grass seedlings were planted per pot with 3-4 replicates per treatment. Inoculated and control pots were irrigated by separate pieces of "aquavlies" irrigation pads from underneath which were kept soaked by wicks from reservoir buckets in a partially shaded, humidified and ventilated glasshouse at 20-30°C. To pots of the grass experiment a modified Hoagland nutrient solution (17), besides trace elements supplying equivalent amounts of 40, 9 and 11 kg·ha^{-1} of N, P and K, respectively, was added after 37 days. Azospirillum populations of roots were determined as described before (14, 15). At suitable time intervals shoot lengths were measured and weigths of tops and roots determined after drying at 70°C.

Sterile seedling experiments: Cereal seeds were sterilized (18) and germinated on sterile water agar at room temperature in the dark. Sterile seedlings with 10-20 mm root lengths were transfered into 18 mm test tubes with 4 ml of sterile N-free Hoagland mineral salt solution (17) made semisolid with 0.2% of Oxoid agar No. 3 and closed by aluminum caps. They were placed on a north side window bench until coleoptiles reached 10-30 mm. For inoculation, *Azospirillum lipoferum* (14) and *A. brasilense* (ATCC 29 710) were grown for 24 h (28°C) in half concentrated nutrient broth, washed with sterile 0.1 m phosphate buffer (pH 7.0), centrifuged and resuspended in sterile buffer to an $O.D._{540}$ of ca. 1.0. 0.1 ml amounts were pipetted onto the agar surface of each tube. After inoculation, culture tubes were placed for 3-4 days in a growth cabinet with a 16 h light (28°C) and 8 h dark (20°C) cycle until the leaves reached the tops of the tubes. Then the caps were replaced by sterile rubber caps, 2 ml of acetylene added and incubation continued under permanent light at 28°C for 24 h. After this, ethylene was determined in 0.5 ml gas samples (14, 15) and nitrogenase activity calculated from total gas volume and incubation time. For maize seedlings, 24 mm sterile test tubes with 10 ml semisolid N-free mineral salt agar, closed with cotton plugs were used. When plants grew higher, they were extended by a second tube, connected by a ring cut from a sterile rubber cap. Shoot length increments during the time of inoculation until determination of C_2H_2 reduction were measured. Illumination was by daylight type incandescent tubes (ca. 2000 lux).

Results and discussion

1. Pot inoculation experiments

In the pot experiment conducted in April and May, stimulations of shoot elongation of 10 or 22% after 22 or 51 d, and of 12 and 9% after 12 or 17 d were observed with inoculated ryegrass and barley plants, respectively (table 1). No such stimulation was found with maize seedlings, although *A. lipoferum* was reported to be more specifically associated with maize in Brazil (19,20). Growth stimulations of other cereals and grasses were often observed after inoculation with *A. brasiliense* (21), which may produce several plant growth promoting substances (22). The growth stimulations observed suggest that such substances may also be produced by *A. lipoferum*. Generally, higher root populations of *Azospirillum* were found with inoculated plants, demonstrating a predominance of the inoculated strain during the first weeks and the possibility to stimulate plant growth promoting substances (table 1).

In the second inoculation experiment with grasses lasting 7 weeks during July and Agust no increases in shoot lengths were found, but dry weights of tops were increased by 14 and 15% with *Arrhenatherum elatius* and *Festuca rubra*, respectively. The shoot: root ratios were increased with *Lolium perenne* and *Festuca rubra*, also indicating a relative stimulation of shoots. During 7 weeks of growth in unsterile soil, *Azospirillum* populations reached also high levels on the roots of control plants, being higher only with inoculated *Lolium perenne* and *Festuca rubra* (table 2). Shoot weight increases without increased shoot lengths are the result of increased tillering such as found with wheat inoculated with *A. brasilense* (23) and may partly also be caused by ultrastructural alterations of the root cortex as found with maize and *Sorghum bicolor* in association with *A. brasilense* (24). Both experiments show that growth stimulations of cereals and grasses can be obtained also by inoculation with a local, temperate

Table 1. Greenhouse experiments with unsterile grassland cereal seedlings inoculated (I) or not inoculated (C) by immersion in suspensions of _Azospirillum lipoferum_ during April and May

Experimental plants	days	mm shoot length C	mm shoot length I	mg dry wt. of tops C	mg dry wt. of tops I	shoot: root ratio C	shoot: root ratio I	log _Azospirillum_ populations$\cdot g^{-1}$ root dry wt. C	log _Azospirillum_ populations$\cdot g^{-1}$ root dry wt. I
Lolium perenne	22	116	128+	17.4	17.8	2.44	2.20	3.77	8.87+
(CV. "NFG")	51	116	142+	122	142	2.21	2.48	7.25	7.88+
Hordeum vulgare	12	133	149+	36	43	2.24	2.12	8.56	n.d.
summer barley	17	212	231+	74	88	1.85	2.12	7.76	9.34+
(CV. "Oriol")									
Zea mays [1]	17	107	101	648 [2]	632	1.10	0.96	4.04	8.07+
(CV. "anjou")									

+ larger than control treatments with $P > 0.05$ or less, as determined by the t-test

1) controls inoculated with heat-killed cells

2) fresh weights

n.d.: not determined

Table 2. Greenhouse experiments with unsterile grass seedlings inoculated (I) or not inoculated (C) by immersion in suspensions of _Azospirillum lipoferum_ during July and August

	days	mm shoot length		mg dry wt. of tops		shoot: root ratio		log _Azospirillum_ populations·g^{-1} root dry wt.	
		C	I	C	I	C	I	C	I
Lolium perenne	45	210	207	2152	2254	1.38	2.30^{+}	8.00	8.75^{+}
Arrhenatherum elatius	49	356	379	2016	2306^{+}	2.01	2.22	8.70	9.13
Festuca rubra	49	290	216	1384	1593^{+}	1.71	2.31^{+}	8.63	9.44^{+}
Dactylis glomerata	49	370	323	3094	3246	2.48	3.07	8.55	8.99

$^{+}$ significant increases, compare table 1

climate isolate of A. lipoferum. However, further experiments have to show which species and cultivars are to be inoculated with what kind of bacterial strains to obtain optimal results from this association.

2. Sterile seedling experiments

Percentages of nitrogenase-active seedlings and mean N_2-ase activities of cereal cultivars after inoculation with A. lipoferum and A. brasilense are summarized in table 3. The very different fractions of N_2-ase positive seedlings of the wheat cultivars Max, Disponent and Kadett after inoculation with A. lipoferum are surprising. They range between 3 and 66% and differ significantly by the chi-sqare test. Similar differences of 20 and 88% might exist between barley cultivars Diana and Koral, however, with five plants tested only from the former this cannot be proved. Also, nearly all maize plants of cv. Mutin were N_2-ase positive. Percentages of active plants were not significantly different after inoculation with A. brasilense in summer wheat Max, summer barley Koral and maize Mutin. Mean N_2-ase activities were 2-4 times higher after inoculation with A. brasilense, possibly because of the lower temperature optimum of the temperate A. lipoferum strain (16). In these monoxenic systems, the property of nitrogen fixation support ("nis") by root exudates may be determined with different bacterial strains, since it seems to depend mainly on the amount and C/N ratio of utilizable root exudates.

The N_2-ase activity within seedlings of the same cultivar follows a log-normal distribution. Plants showing the same activity when placed in a rank order, the logarithms of their N_2-ase activities follow a linear function as shown by wheat cv. Max inoculated with A. lipoferum (fig. 1). Thus, with this test, from a set of N_2-ase positive seedlings one may always select some with a high activity which seem to express the complex phenotype of "nis" in an optimal way. Large genotypic variations in associative N_2-ase activity were also observed with maize hybrids (25,12), with rice

Table 3. Nitrogenase (C_2H_2) activity (nmol $C_2H_4 \cdot h^{-1} \cdot plant^{-1}$) of sterilized cereal seedlings in mineral salt agar, inoculated with Azospirillum lipoferum or A. brasilense

Cultivars	plants tested	% of active plants [1)] lip.	bras.	mean activity ± S.E. lip.	bras.
Summer wheat "Max"	29	66_a	86_a	8.4 ± 0.2	26.0 ± 3.4
"	63	63.5_a	n.d.	9.4 ± 1.4	n.d.
Winter wheat "Disponent"	5	0	40	0	1.2 - 21.3
"	49	26.5_b	n.d.	2.7 ± 1.1	n.d.
Summer wheat "Kadett"	30	3.3_c	n.d.	(7.3)	n.d.
Summer barley "Koral"	25	88_a	84_a	15.7 ± 4.3	33.3 ± 1.5
Winter barley "Diana"	5	20	20	(61.2)	(11.5)
Summer rye "Teku"	5	20	40	(11.5)	1.1 - 43.2
Winter rye "Kustro"	5	0	20	0	(2.4)
Maize "Mutin"	9	89_a	100_a	13.5 ± 5.1	70.8 ± 18.3

1) lip. = A. lipoferum, bras. = A. brasilense

n.d. = not determined, in brackets: single values only, percentage values followed by different letters differ significantly by P > 0.05 or less, as determined by the chi-square test

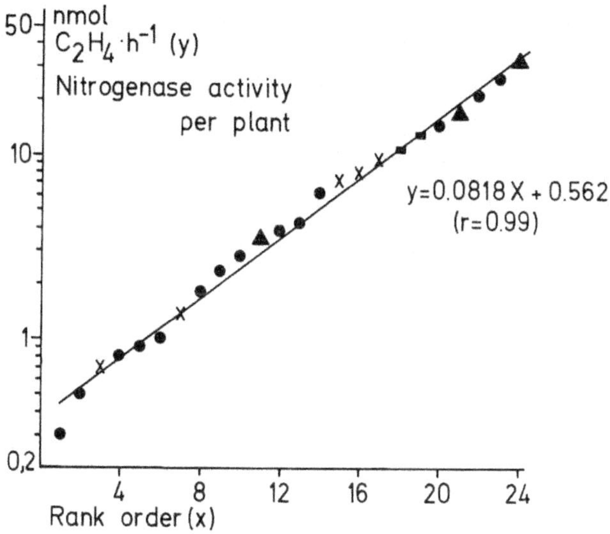

Figure 1. Nitrogenase activity of sterilized seedlings of summer wheat cv. Max after inoculation with A. lipoferum: Crosses, triangles and squares are averages of 2,3 or 4 equal plants

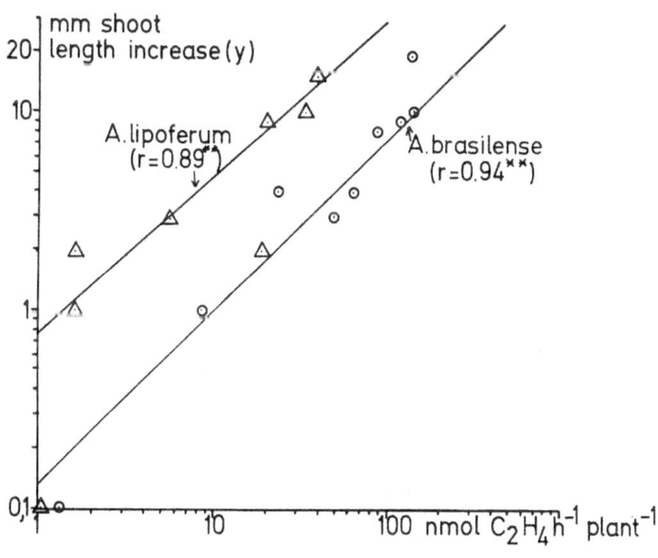

Figure 2. Relationship between N_2-ase activity of sterilized maize seedlings and shoot length increments after inoculation with A. lipoferum and A. brasilense

(11) and with wheat cultivars (3). There was no correlation between shoot growth during the test and N_2-ase activity with small grained cereals. However, with N_2-ase positive maize seedlings, a linear relationship could be shown between the logarithms of shoot length increase (growth vigour) and the logarithm of N_2-ase activity (fig. 2).

This seedling test seems to be promising to screen for the character of "nis" for plant breeding purposes. Since cultivars never have been selected for "nis" by modern breeding they are bound to be different and inhomogenous in this respect. It still has to be demonstrated, however, whether high N_2-ase activities in seedlings are also correlated with higher activities at later growth stages, with yield increases after inoculation and with high activities in subsequent seedling generations.

Acknowledgments:
The author thanks the Deutsche Forschungsgemeinschaft for support of part of this work, Dr. I. Nioh for help in the pot experiment with grasses, Frau Kirsten Barg and Frau Susanne Scheil for their skilled technical assistance.

References

1. Subba Rao, N.S., Tilak, V.B.R., Sing, C.S. and Lakshmi-Kumari (1979). Current Sci. 48, 133-134

2. Schank, S.C., Weier, K.L. and MacRae, I.C. (1981). Appl. Envir. Microbiol. 41, 342-345

3. Avivi, Y. and Feldman, M. (1982). Israel J. Bot. 31, 237-245

4. Bouton, J.H., Smith, R.L., Schank, S.C., Burton, G.W., Tyler, M.E., Littell, R.C., Gallagher, R.N. and Quesenberry, K.H. (1979). Crop. Sci. 19, 12-16

5. Taylor, R.W. (1979). Trop. Agric. 56, 361-365

6. Smith, R.L., Schank, S.C., Bouton, J.H. and Quesenberry, K.H. (1978). Ecol. Bull. (Stockhom) 26, 380-385

7. Kapulnik, Y., Sarig, S., Nur, I., Okon, Y., Kigel, J. and Henis, Y. (1981). Expl. Agric. 17, 179-187

8. Kapulnik, Y. Sarig, S., Nur, I. and Okon, Y. (1982). Israel J. Botany 31, 247-255

9. Singh, S., Ganguly, T.K., Neelakantan, S., Singh, K., Singh, A. and Tomer, P.S. (1980). Ind. J. Agric. Sci. 50, 721-725

10. Rai, S.N. and Gaur, A.C. (1982). Plant and Soil 69, 233-238

11. Sano, Y., Fuji, T., Hirota, Y. and Komagata, K. (1981). Crop. Sci. 21, 758-760

12. Ela, St.W., Anderson, M.A. and Brill, W.J. (1982). Plant Physiol. 70, 1564-1568

13. Rennie, R.J., de Freitas, J.R., Ruschel, A.P. and Vose, P.V. (1983). Can. J. Botany 61, 1667-1671

14. Jagnow, G. (1981). In W. Klingmüller (ed.): Azospirillum I, Experientia Suppl. Vol. 42, 100-107

15. Jagnow, G. (1983). Z. Pflanzenernähr. Bodenkd. 146, 217-227

16. Jagnow, G. (1983). In W. Klingmüller (ed.): Azospirillum II, Experientia, Suppl. Vol. 48, 127-137

17. Robbins, W.J. and Kavanagh, F. (1938). Amer. J. Bot. 25, 229-236

18. Barber, D.A. (1967). J. exp. Botany 18, 163-169

19. Baldani, V.L.D. and Döbereiner, J. (1980). Soil Biol. Biochem. 12, 433-439

20. Döbereiner, J. and Baldani, V.L.D. (1979). Can. J. Microbiol. 25, 1264-1269

21. Okon, Y. (1984). In C. Veeger and W.E. Newton (eds.): Advances in nitrogen fixation research. Martinus Nijhoff, The Hague and Dr. W. Junk, Wageningen, pp. 303-309

22. Tien, T.M., Gaskins, M.H. and Hubbell, D.H. (1979). Appl. Envir. Microbiol. 37, 1016-1024

23. Kapulnik, Y., Kigel, J., Okon, Y. and Henis, Y. (1981). Plant and Soil 61, 65-70

24. Lin, W., Okon, Y. and Hardy, R.W.F. (1983). Appl. Envir. Microbiol. 45, 1775-1779

25. von Bülow, J.F.W. and Döbereiner, J. (1975). Proc. Natl. Acad. Sci. USA 72, 2389-2393

RESPONSE OF PLANT ROOTS TO INOCULATION WITH AZOSPIRILLUM BRASILENSE AND TO APPLICATION OF INDOLE ACETIC ACID

W. Kolb and P. Martin
Institut für Pflanzenernährung, Universität Hohenheim, Postfach 700562, D-7000 Stuttgart 70, FRG

Summary: By inoculation with *Azospirillum brasilense* Ft 326 root elongation of *Beta vulgaris* ssp. *cicla* was stimulated and the number of the lateral roots was increased. The strain used is a mutant which produces more auxin than the wild strain Sp Cd from which it was obtained by Hartmann (Bayreuth). In these experiments the plants were cultivated in boxes filled with soil which were placed in an inclined position so that root growth was along a plexiglass plate which could be removed shortly for spraying. Application of IAA to wheat plants grown in the same way or to plants grown under sterile conditions in agar plates led to similar results. The dominating phytohormone determined by radioimmunoassay in the supernatant of the medium of *Azospirillum* was IAA. ABA was found in moderate amounts and cytokinins in traces only.

Introduction

Numbers of N_2-fixing bacteria in the rhizosphere usually represent only a very small proportion of the total population of rhizosphere bacteria (4, 6). Inoculation may increase the numbers of N_2-fixing bacteria in the rhizosphere and stimulate possible beneficial effects on plant growth by nitrogen fixation. However, in the past few years there have been increasing indications that besides N_2-fixation, liberation of phytohormones by the N_2-fixing bacteria, particularly Azospirilla, may play an important role. Changes in root morphology after inoculation of *Azospirillum* were repeatedly observed (3, 5, 8, 9, 10). One possible explanation for a beneficial effect of hormones may be seen in an increase in the nutrient absorbing root surface.

Azospirillum III: Genetics, Physiology, Ecology
Ed. by W. Klingmüller

In the present paper initial results are presented from investigations on the effect of inoculation with *Azospirillum brasilense* and of direct application of IAA (indole acetic acid) onto the roots on root elongation and the number of lateral roots and root hairs.

Materials and Methods

For experiments with soil as growth medium (unsterile conditions) *Beta vulgaris* ssp.*cicla* and *Triticum aestivum* (spring wheat cv. "Schirokko") were used, for experiments with nutrient agar as medium (sterile conditions) wheat only.

For the cultivation of plants in soil plastic boxes were used (400 x 200 x 100 mm) which contained 5 kg of a loamy field soil sieved to 2 mm with the addition of 15 % (weight) vermiculite. Fertilizers were not added. After planting the seedlings on the upper small side of the boxes, the boxes were placed in an incline position in such a way that the roots grew along one side which was covered with a plexiglass plate. The plate was removed shortly for spraying bacteria suspensions or IAA solutions 7 to 9 days after germination. *Azospirillum brasilense* Ft 326 was used for inoculation. This strain is a mutant which produces more auxin than the wild type Sp Cd from which it was obtained by Hartmann et al. (1). Bacteria for inoculation were grown in CASO-Bouillon (Merck). Control plants were treated with autoclaved bacterial cultures.

Sterile cultivation of plants was performed in petri dishes (200 mm). Seeds were desinfected and precultivated on sterile agar. After 2 days seedlings were transferred to petri dishes which contained agar (0.8 %, 7 mm layer) with the following nutrient composition (mM/l): NH_4NO_3 0.05, KCl 0.5, KH_2PO_4 0.064 $MgSO_4$ 0.3, $CaCl_2$ 0.25. Adequate amounts of micronutrients were added. For cultivation the plates with 3 plants each were placed vertically. Roots were protected against light with aluminium foil.

IAA-solutions sterilized by filtration were sprayed evenly onto the agar 2 days after transferring the seedlings to the plates. All plants were grown in a controlled environment chamber with a 16 h day/8 h night cycle at 21°C.

For the determination of hormones produced by *A. brasilense* Ft 326 bacteria were cultivated in a liquid medium with a nutrient composition according to the medium described by Rennie (7). 0.1 g NH_4NO_3/l was also added. Tryptophan was not added. Analysis was carried out by radioimmunoassay.

Results and Discussion

The method used for the experiments with plants growing in soil allowed an even spraying of the inoculum onto the root and the soil surface along which the roots grew. Root growth could

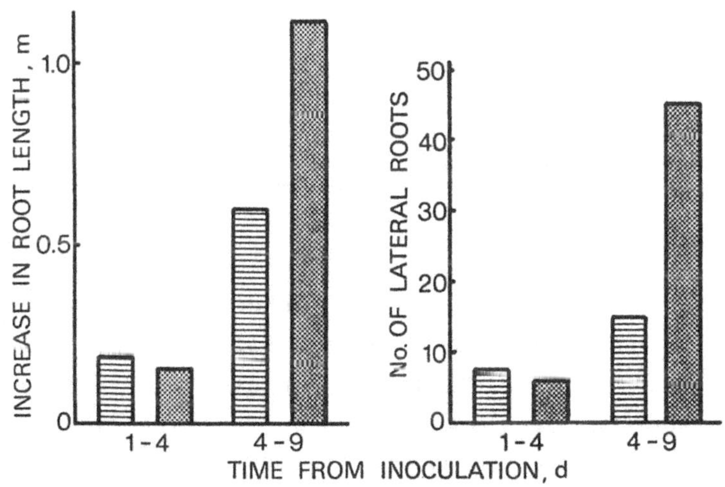

Figure 1. Response of roots of *Beta vulgaris* ssp. *cicla* to inoculation with *A. brasilense* Ft 326. The inoculum was sprayed 9 d after sowing onto the rooted side of a soil containing box, which was placed at an incline during the growth period (Increase in root length and increase in number of lateral roots significant at 5% level).

▤ Control ▩ Inoculated

be followed along the plexiglass plate. Fig. 1 shows that root elongation was stimulated by inoculation with *A. brasilense*. Following inoculation the effect was visible after the 4th day. The number of lateral roots was also significantly increased.

The strain of *A. brasilense* used is a mutant of Sp Cd with a higher production of auxin than the wild type. Measurements of hormones in the liquid growth medium by radioimmunoassay confirmed the high ability to produce auxin. Tab. 1 shows that besides IAA ABA was also present in the medium but only traces of cytokinins were found (Gibberellins were not determined).

Table 1. Phytohormones in the supernatant of the medium of *A. brasilense* Ft 326 after 60 d cultivation (10^8 bacteria/ml)

Phytohormone	conc.ng ml^{-1}
Indole acetic acid	1.5
Abscisic acid	0.4
trans-Zeatin/zeatin riboside	traces
Isopentenyladenine/-adenosine	traces

The relatively high amount of IAA found in the growth medium of *Azospirillum* suggested that the main effect on root morphology is caused by this hormone. Therefore, an attempt was made to mimick the effect of the bacterium on root growth by a direct application of IAA onto the roots. Fig. 2 shows a stimulation of root elongation of wheat plants when a concentration of 10^{-9} IAA/l was applied. The number of lateral roots was slightly (but not significantly) increased.

Since in soil exogenously applied auxin is readily subjected to decomposition, root growth can only be affected for a limited time. This could explain the fact that not all experiments of this type gave clear results. Therefore, auxin applications were

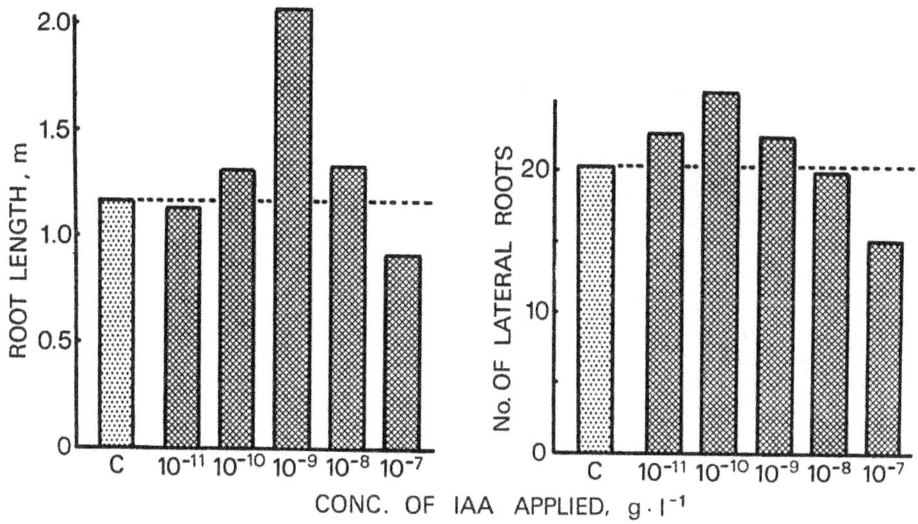

Figure 2. Response of roots of wheat plants to application of IAA. Solutions were sprayed 7 d after sowing onto the rooted side of a soil containing box which was placed at an incline during the growth period. Measurements were made 7 d after the treatment (Increase of root length at a conc. of 10^{-9} significant at 5% level).

Control IAA application

repeated with wheat plants cultivated under sterile conditions. For this purpose agar plates were used. The effect of IAA on root elongation was similar to that of plants grown in unsterile soil (Fig. 3). There was also a small stimulating effect on the number of lateral roots (results not given in Fig. 3). Countings of the number of root hairs which were carried out in this experiment showed a small increase at optimal IAA concentrations and a large increase at a concentration which is inhibiting for root elongation. The higher density in the latter case can be explained by an inhibition of cell elongation and therefore a higher cell density per unit root length.

The comparison of the response of root development to inoculation with *A. brasilense* and to application of pure solutions of IAA suggests that the causing agent produced by the bacteria is

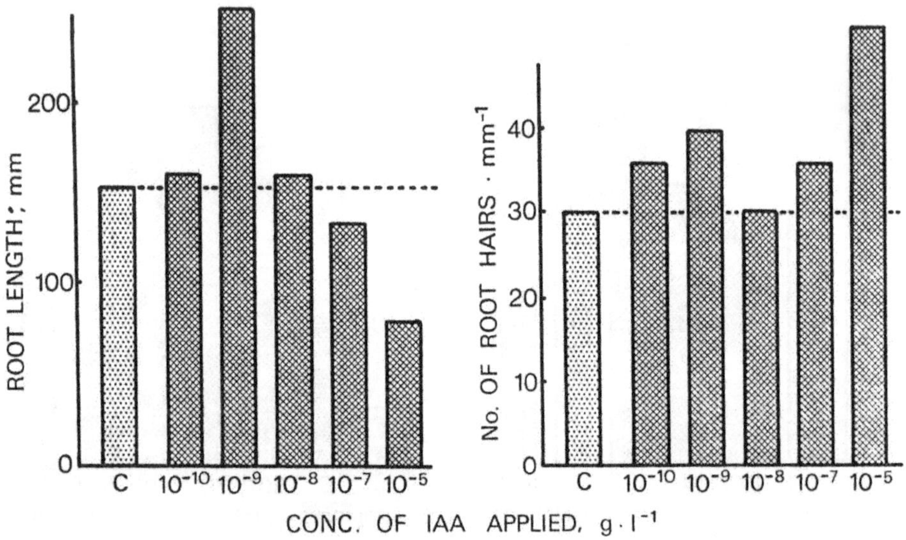

Figure 3. Response of roots of wheat plants to application of IAA. IAA solutions were sprayed onto agar plates with 4 d old plants. Measurements were made 4 d after application (Increase in root length at a conc. of 10^{-9} and increase in number of root hairs at a conc. of 10^{-5} significant at 5% level).
Control IAA application

IAA. A similar comparison in regard to the action of IAA on roots was made by Jain and Patriquin (2). These authors identified a substance produced by *Azospirillum* causing branching of wheat root hairs as IAA and could mimick the branching effect by pure solutions of IAA.

The results support the view that inoculation with *Azospirillum* may promote plant growth by the production of hormones.

Acknowledgements

We thank Prof. Dr. F. Bangerth, Institut für Obst-, Gemüse- und Weinbau Universität Hohenheim, for providing facilities in his laboratory to carry out analysis on hormones by radioimmunoassay. We are also grateful to Dr. A. Hartmann, Institut für Genetik, Bayreuth, for providing *A. brasilense* Ft 326. The work has been supported by the Deutsche Forschungsgemeinschaft.

References

1. Hartmann, A., Fußeder, A., Klingmüller, W. (1983). In *Azospirillum,* Genetics Physiology Ecology (Ed. Klingmüller, W.).Experientia Suppl. 48, 78-88

2. Jain, D.K. and Patriquin, D.G. (1985). Can. J. Microbiol. 31, 206-210

3. Kapulnik, Y., Gafny, R. and Okon, Y. (1985). Can. J. Bot. 63, 627-631

4. Kolb, W. (1984). Doctoral thesis, Universität Hohenheim

5. Martin, P. and Glatzle, A. (1982). In *Azospirillum*, Genetics Physiology Ecology (Ed. Klingmüller, W.). Experientia Suppl. 42, 108-120

6. Okon, Y., Albrecht, S.L. and Burris, R.H. (1977). Appl. Environ. Microbiol. 33, 85-88

7. Rennie, R.J. (1981). Can. J. Microbiol. 27, 8-14

8. Rinaudo, G. (1983). Fixation hétérotrophe de l'azote dans la rhizosphere du riz. Orstom, Paris. I.S.B.N. 2-7099-0704-6

9. Tien, T.M., Gaskins, M.H. and Hubbell, D.H. (1979). Appl. Environ. Microbiol. 37, 1016-1024

10. Venkateswarlu, B. and Rao, A.V. (1983). Plant Soil 74, 379-386

ELECTRON MICROSCOPY OF YOUNG WHEAT ROOTS INOCULATED WITH AZOSPIRILLUM

J. H. Whallon[1], G. Acker[2] and H. El-Khawas
Lehrstuhl für Genetik and [2] Electron Microscopy Laboratory
Universität Bayreuth, Universitätsstraße 30
D-8580 Bayreuth, FRG

Summary: Germinated seeds of *Triticum aestivum* L. were inoculated with *Azospirillum brasilense*, and one to two weeks later were prepared for electron microscopy. SEM revealed bacterial growth varying from sparse to heavy on all surfaces of all roots, with growth on the roots hairs being sparse near the root tip and sparse to moderately heavy near the seed. Some cells and some intercellular spaces contained a few bacteria, but there was no gross infestation. TEM confirmed the presence of a few bacteria inside the root, and revealed that a zone relatively free of bacteria separated the majority of the organisms from the surface of the root. Keywords: *Azospirillum*, wheat, electron microscopy, roots.

Introduction

Inoculation of grasses with *Azospirillum*, even under conditions thought to be most favorable, frequently has not produced the expected benefits. Ultrastructural investigations may help us to understand why the effects of grass-diazotroph associations are so variable. This paper presents the results of a preliminary study undertaken to explore the interaction between *Azospirillum* and developing wheat roots.

---

[1] Corresponding address: Department of Crop and Soil Sciences, Michigan State University, East Lansing, MI 48824 USA

Azospirillum III: Genetics, Physiology, Ecology
Ed. by W. Klingmüller

Materials and Methods

Wheat (*Triticum aestivum* L. var. Scirocco) seeds were surface sterilized with 0.1% $HgCl_2$ and germinated under aseptic conditions in the dark at 20° C. After 5-8 days, 28 seedlings that showed no signs of contamination were transferred to test tubes containing agar or sterilized vermiculite plus Knop's nutrient medium without nitrogen. Following transfer, 20 of the test tubes were inoculated with *Azospirillum brasilense* 29710 (Cd) from a log phase culture with a concentration of 10^8 organisms per ml; 0.375 to 0.5 ml of inoculum was used for each 10 ml of agar or vermiculite in each tube. The seedlings were subsequently maintained in a growth chamber with a light and temperature cycle of 11-hours on (20° C)/13-hours off (14° C). The root area of each agar tube was shielded from light with aluminum foil.

Seven to 14 days after inoculation the plants were harvested and small pieces of root were prepared for electron microscopy by fixation in 2% glutaraldehyde for 2-4 hours. For scanning electron microscopy, the samples were taken from the distal 2 mm of the root (root tip, root cap, zones of elongation and differentiation, youngest mature root hairs) and from the proximal 10 mm (area adjacent to the seed, lateral roots, oldest mature root hairs). Following fixation the samples were rinsed twice in phosphate buffer, dehydrated in an ethanol series, critical-point dried from CO_2, coated with gold, and examined in a Novascan 30 or JEOL 35C scanning electron microscope operating at 15 KV. For transmission electron microscopy samples were fixed as described previously (7) and embedded in Spurr's resin. Thin sections were stained with uranyl acetate/lead citrate (6) and examined in a Zeiss 109 electron microscope at 80 KV.

Results

Examination of the exterior of the samples revealed bacteria on all areas of all inoculated roots, although the amount varied greatly from place to place on each root, and from root to root.

Figures 1 and 2 show heavy, multi-layer growth on a 7-day root grown in agar and moderate growth on a 14-day root grown in vermiculite, respectively. The pleiomorphic nature of the bacteria is apparent, especially in Fig. 2. In samples taken from the younger part of the roots (i.e., the distal region), root hairs had relatively few bacteria on them (Fig. 3), even in areas where the surface was very heavily colonized. By contrast, in the older regions, adjacent to the seed, there was moderately heavy growth on some root hairs (Fig. 4). This was observed only in vermiculite-grown roots, and surprisingly, was not accompanied by equally heavy growth on the surface of the same sample. Bacteria were also found on the lateral roots, and were well established even on the lateral root caps.

Several samples were fractured to permit observation of the root interior. Fractured roots displayed no gross infestation of any cells, nor of any of the large intercellular spaces in the cortex of older root regions. However, a few bacteria were seen in some of the small intercellular spaces of the cortex, and in the vacuolar region of some cells (Figs. 5, 6).

Samples from the mid-region of a single root were examined by transmission electron microscopy. This revealed a highly pleiomorphic bacterial population, with wide variation in size, shape, degree of encapsulation and presence of inclusion bodies, presumably containing poly-ß-hydroxybutyrate (Figs. 7-10). The larger organisms in Fig. 2 may thus be those with a large amount of PHB and a well developed capsule. Some bacteria were in contact with the root surface, in both random and polar

---

Figs. 1-6. Scanning electron micrographs of root surface and interior. Fig. 1. Multi-layer colonization of root grown in agar. Fig. 2. Bacteria of varying sizes in mucigel of root grown in vermiculite. Fig. 3. Bacteria on root hairs near root tip. Fig. 4. Bacteria on root hairs near seed. Fig. 5. Part of root fractured transversely showing bacteria in vacuolar region of two adjacent cells (arrowheads). Note also that the root surface is heavily colonized but that the root hairs are not. Fig. 6. Higher magnification of one of the cells in Fig. 5. B, bacteria; C, cortex; E, epidermis; R, root hair. Bar=10 um in Figs. 1,3,4,5, 1.0 um in Figs. 2,6.

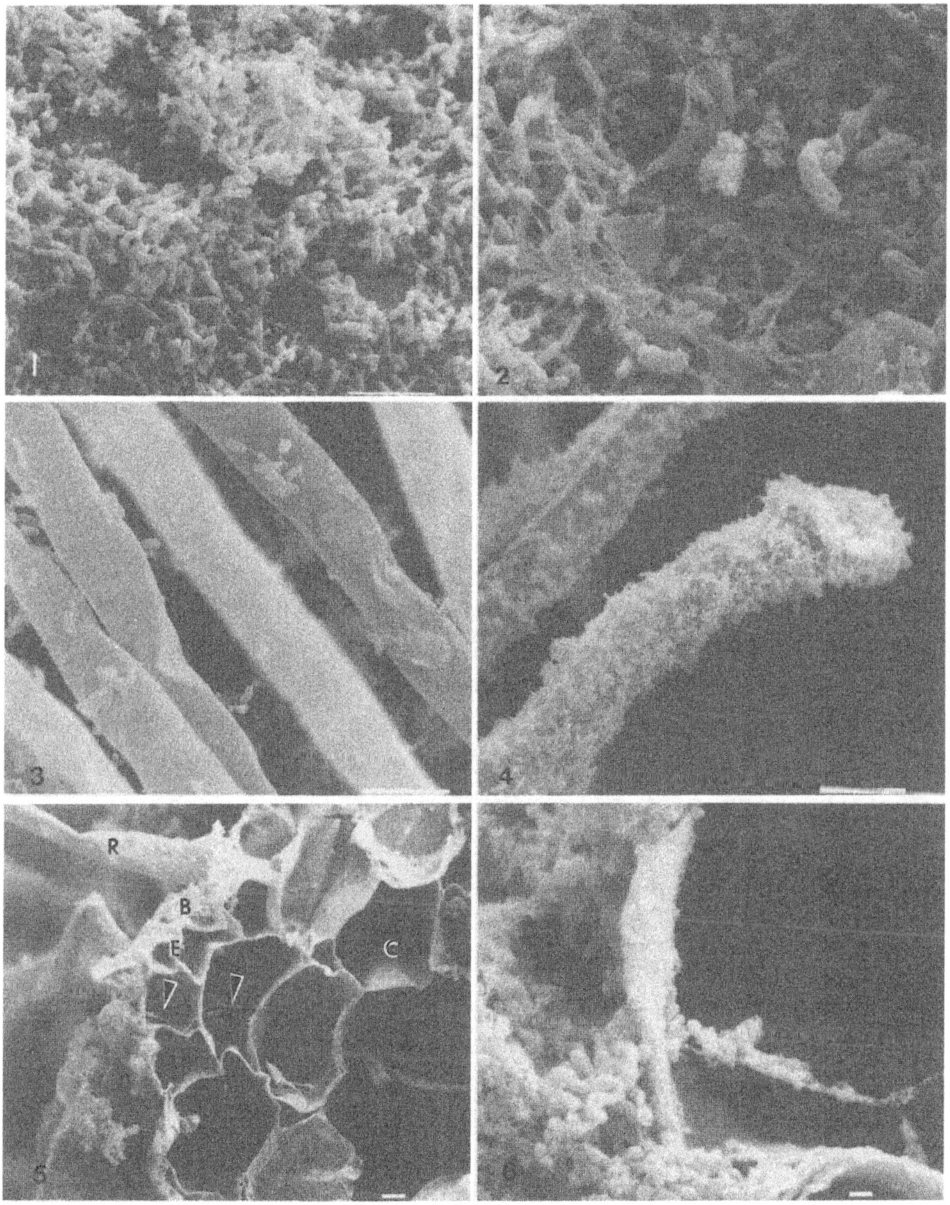
1
2
3
4
R
B
E
C
5
6

orientation (Figs. 7,8,10). However, the majority of the bacteria were located some distance away from the root (Figs. 9,10), separated from it by a 5 to 15 um space containing few or no bacteria, and little or no other structural substance (e.g., mucigel) except agar.

The outer epidermal wall of the root was marked by a number of unevenly distributed translucent areas (Figs. 7,8,10), and by patches of what appeared to be a polysaccharide-type coating. As indicated by their lack of cytoplasm, some of the epidermal cells were dead, and a few of these dead cells contained a single bacterium each (Fig. 10).

Discussion

Our observations suggest that the bacteria are differentially attracted to the root surface and to the root hairs; that the attraction of the root surface for the bacteria does not change between younger and older regions of the same root, but that that of the root hairs may change, at least in vermiculite-grown roots. As an explanation of the relatively heavy growth of bacteria on some root hairs, cell death followed by release of breakdown products is certainly a possibility, considering the age and location of the root hairs. Although the bacterial population in these samples was highly pleiomorphic, neither on the root surface nor on the root hairs did we find anything comparable to the filled sperical bags that Patriquin, Döbereiner and Jain (5) reported seeing on young wheat roots.

We know from the work of Foster and Rovira (3) that bacteria

Figs. 7-10. Transmission electron micrographs through midsection of a heavily colonized root. Fig. 7. Pleiomorphic bacteria in random contact with root surface. Fig. 8. Perpendicular orientation of one bacterium to root surface. Fig. 9. Bacteria lying close to root surface, but separated from it by a distance of about 4 um. Fig. 10. Overview of root and bacteria. Very few bacteria make contact with root, except in the vicinity of the cell which already contains a single bacterium (arrowhead). C, bacterium with capsule; CW, cell wall of epidermal cell; E, epidermal cell; P, poly-ß-hydroxybutyrate inclusion.

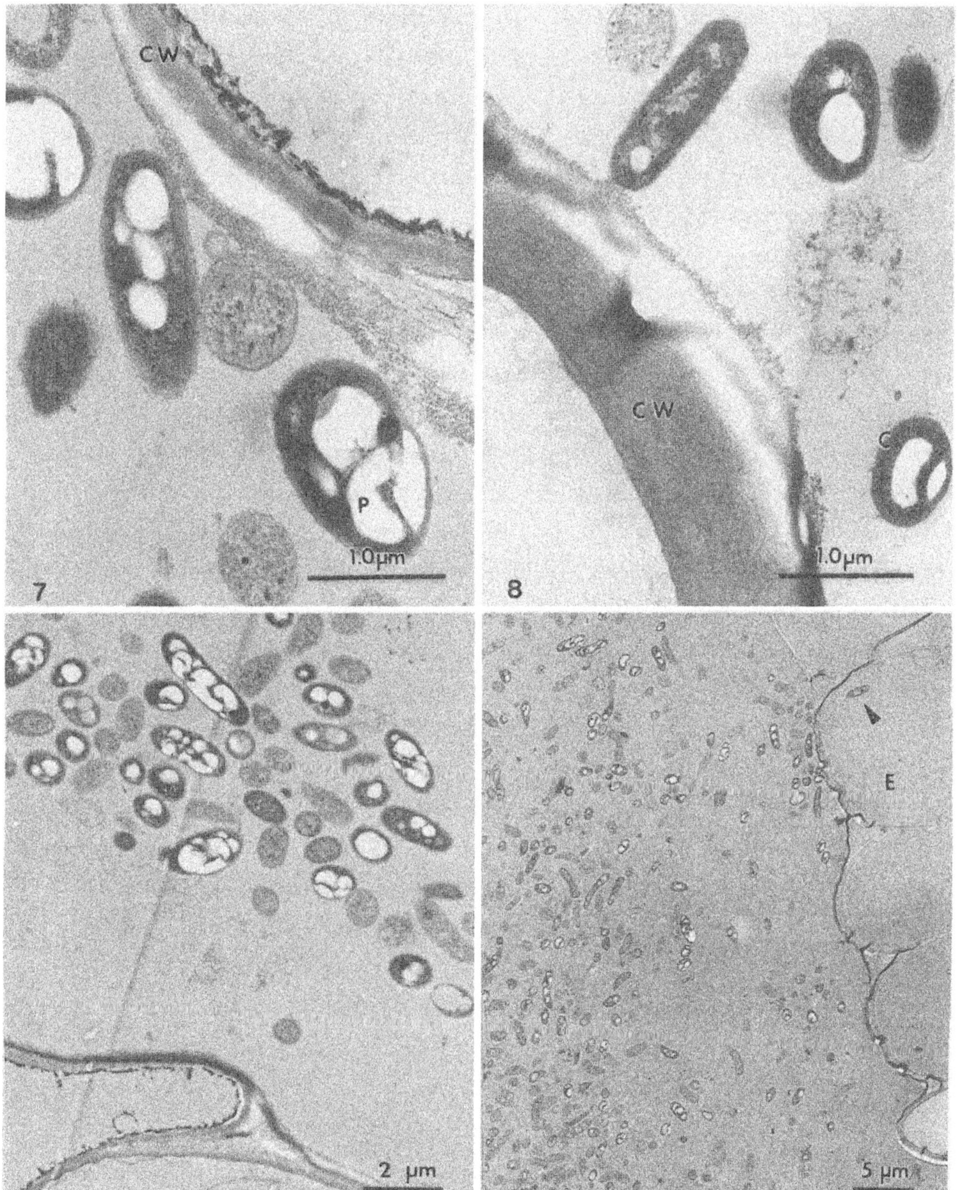
CW
P
1.0μm
7
CW
C
1.0μm
8
2 μm
9
E
E
5 μm
10

do enter dead cells in older wheat roots in significant numbers. Whether or not they enter living cells is a more controversial matter, and some authors (1,5) have expressed the view that invasion of healthy cells cannot occur. Although we saw no bacteria in living cells and no gross infestation, it is important to remember that none of these plants was more than three weeks old.

The fact that the majority of the bacteria were lying some distance from the root surface, if not an artifact, is intriguing since it suggests an explanation for the low rate of invasion of the root - namely, that relatively few bacteria ever come in contact with it. No such separation phenomenon was reported in the studies of Azospirillum-sugar cane callus association by Berg et al. (see their Fig. 10), but Marshall et al. (4) have described for marine bacteria the existence of an energy barrier between the negatively charged plant surface and the negatively charged bacterial surface. Dazzo et al. (2) have noted that, in the case of Rhizobium, flagella are important in helping the bacteria to bridge this barrier. In this context, the high degree of pleiomorphism in Azospirillum may be important. In the SEM, organisms having the size and shape of the one shown in Fig. 8 often appeared closer to the root surface than the (presumably) encapsulated forms, and to be the form involved in monolayer attachments. Whether or not this is the only form that makes polar contact with the surface, and whether conditions that increase the prevalence of this (or some other) form are the same as those in which Azospirillum has a beneficial effect on plant growth, remain to be determined.

References

1. Berg, H., Vasil, V. and Vasil, I. K. (1979) Protoplasma 101, 143-163.
2. Dazzo, F. B., Truchet, G., Sherwood, J., Hrabak, E., Abe, M. and Pankratz, S. (1984) App. Env. Micro. 48, 1140-1150.

3. Foster, R. C. and Rovira, A. D. (1976) New Phytol. 76, 343-352.
4. Marshall, K. C., Stout, R. and Mitchell, R. (1971) J. Gen. Micro. 68, 337-348.
5. Patriquin, D. G., Döbereiner, J. and Jain, D. K. (1983) Can. J. Micro. 29, 900-915.
6. Reynolds, E. S. (1963) J. Cell Biol. 17, 208-212.
7. Traub, W. H., Acker, G. and Kleber, J. (1976) Chemotherapy 22, 104-113.

CYSTS OF AZOSPIRILLA UNDER VARIOUS CULTURAL CONDITIONS

Lakshmi Sadasivan and Carlos A. Neyra
Department of Biochemistry and Microbiology, Cook College
Rutgers University, New Brunswick, NJ 08903

Summary: Cyst formation and germination in *A. brasilense* ATCC 29145 and *A. lipoferum* ATCC 29707 was observed by using routine laboratory staining and phase contrast and electron microscopy. Different cultural conditions and carbon sources were tested for the induction of cyst formation. Various morphological forms were found to be resistant to desiccation for a period of more than six months. Even though morphologically and ultrastructurally, *A. brasilense* cysts obtained under nitrogen limiting conditions resemble *Azotobacter vinelandii* cysts in some respects, the exact sequence of events associated with encystation and germination is yet to be understood. Unlike *A. vinelandii*, *A. brasilense* is found to synthesize and excrete brown-black melanin-like granular pigments during encystation under nitrogen limiting conditions. An attempt to separate the different cyst forms by ultracentrifugation in a sucrose density gradient, yielded three different bands. Most of the dense cyst forms with aggregated pigmented granules were pelleted at the bottom of the 70% gradient.

A desiccation resistant form in *A. lipoferum* with abundant PHB granules, was obtained in a liquid medium containing fructose and KNO_3, but the formation of a definite cyst-like structure and melanin-like granules were not observed in either solid or semisolid medium. Thus the formation of cysts and melanin pigments appear to be controlled by genetic and envirmental factors.

Keywords: Cysts, exine, intine, desiccation resistance, germination, brown-black pigments, melanin, sucrose density gradient.

Azospirillum III: Genetics, Physiology, Ecology
Ed. by W. Klingmüller

Introduction

Metabolically dormant cells with considerable resistance to deleterious physical conditions, particularly to desiccation, are designated as cysts (13, 14). Azotobacter are the only free-living N_2 fixing bacteria in which the process of cyst formation and germination have been thoroughly studied (10, 11, 14, 18). Azotobacter cells exposed to certain carbon sources, under nitrogen limiting conditions, are induced to form cysts. During the process, vegetative cells undergo a series of metabolic and morphological changes (7, 10, 16). The protoplast shrinks and forms a central body with the development of Poly-β-hydroxybutyric acid and the outer exine or exine related material is built from the blebs of lipoproteins formed on the outer surface of the central body (4, 13, 16). These structures are believed to confer resistance to various deleterious conditions and are likely responsible for the reported long-term survival of Azotobacter in soil stored under laboratory conditions for many years (17). Azotobacter are also known to form giant cysts or cysts with multiple central bodies during germination (1, 2).

In the case of azospirilla, only few reports of cyst formation are available. They were based on the criterion of desiccation resistance by cell forms which appeared modified from the normal vegetative cells (6, 9). Neither of these reports (6, 9), however, presented any ultrastructural details of the probable various components of a typical cyst.

In this paper, we present in Azospirillum brasilense some of the morphogenetic events occurring during cyst-formation and germination as observed under phase contrast and electron microscopy. A discussion of the similarities with Azotobacter cysts is also included.

Materials and Methods

Bacterial Strains and Cultural Conditions: Azospirillum brasilense sp7 (ATCC 29145) and A. lipoferum sp59b (ATCC 29707) were used for this study. These strains were routinely subcultured on nutrient agar (Difco) slants and maintained at 30°C.

Minimal salts medium described as medium A by Neyra and Van Berkum (8) was used as the basic medium with modifications to incorporate different carbon sources. Fructose or malate at a concentration of 8 mM and 40 mM or n-Butanol (0.3%) were added individually as carbon sources for the induction of encystation. Three different cultural conditions were employed which were liquid, solid and semisolid, the latter was prepared with either 0.05% or 0.175% Difco agar. In liquid and solid medium, a nitrogen source in the form of 0.5 mM KNO_3 was added.

A log phase culture from nutrient broth ($O.D._{660}$ = 0.8) was washed three times in $K\text{-}PO_4$ buffer (pH=6.8) and finally suspended in the buffer to give an O.D. = 0.8. A loopfull was streaked on solid agar plates, whereas 0.1 ml inoculum was added to 100 ml of semisolid and 10 ml in 100 ml liquid medium. The solid plates and the flasks with semisolid medium were kept stationary while the liquid medium cultures were continuously shaken at 200 rpm in a gyrotary shaker. All plates and flasks were incubated at 34°C in two sets under continuous light and darkness for at least one week.

The growth obtained in all three cultural conditions were allowed to desiccate naturally in their respective medium and after drying, the growth in liquid or semisolid medium and colonies on solid agar were observed under a phase contrast microscope either unstained or stained according to Vela and Wyss (18).

The desiccated growth were also fixed in 1% glutaraldehyde, stained either in uranyl acetate or ruthenium red according to Cole and Popkin (3), and processed for electron microscopy. The negative staining preparations and ultrathin sections were then observed under a transmission electromicroscope (Siemens Model Elmiscope 1A).

Separation by Sucrose Density Gradients

Gradients of sucrose in water (55%, 60%, 65% and 70% wt/wt) were prepared. Each concentration was carefully layered on top of one another with 70% at the bottom and 55% in the top. The brown pellicle obtained in the semisolid medium was carefully removed with a pasteur pipette, suspended in 100 mM K-PO_4 buffer (pH 6.8) and was layered on top of the 55% gradient. It was then centrifuged at 20°C for 4 hours at 40 K rpm in a Beckman L5-75B ultracentrifuge. The cells and cysts separated into bands were photographed later under electronmicroscope after negative staining.

Characterization of Pigments

The brown-black granular pigments formed in the colonies and pellicles were tested for their solubility in water, organic solvents like benzene, ethanol, chloroform or acetone, cold or hot alkali (0.5 N NaOH or KOH) and their bleaching properties in 10% hydrogen peroxide or commercial chlorox according to Ivis and Holmes (5). The absorption spectrum of the 0.5 N KOH (hot) soluble pigment was recorded on a Cary IFD spectrophotometer between 280 and 400 nm.

Results

Induction of Cyst Formation: Butanol used as a sole carbon source did not support growth of either *A. brasilense* or *A. lipoferum* under any cultural conditions. Media containing 8 or 40 mM fructose and only the 40 mM malate yielded encysting cells in *A. brasilense* associated with the formation of brown pigmented colonies on solid agar plates and a brown pellicle in semisolid medium. *A. lipoferum* did not show any development of brown pigments under any of the three cultural conditions instead the medium turned slightly yellowish and the cells under the phase contrast microscope, were non-motile and highly pleomorphic with most of the cells noticeably enlarged. Figure 1 shows the cells from a dry pellicle of *A. lipoferum* under phase contrast microscope and Figure 2 is the ultrathin section of the same in which no poly β-hydroxybutyrate granules are present and the protoplasm is seen shrunken from the outer wall.

The use of fructose as a carbon source appears to be ideal for the early induction of cyst formation (within 48 to 72 hours) in *A. brasilense*. Malate, at low concentrations did not yield any cysts but at higher concentration (40 mM) and after a week of incubation, a brown pigmentation develops and the cyst formation occurs. The brown to black pigmentation develops quicker in the light incubated flasks as compared to dark incubation Cytological observation of the brown colonies or pellicle from fructose grown A. brasilense showed round cyst-like cells with a hard coat and a dense internal region as shown in Figure 3. The brown colonies after desiccation for 6 months could germinate and shed the outer coat. Figure 4 shows cysts in various stages of germination and the horseshoe shaped residual coats after the vegetative cell emerges out.

Ultrastructure of the Encysting A. brasilense

Figure 5 shows the ultrathin section of an encysting cell prior to complete desiccation of a brown colony. The shrinking of the protoplasm is evident with blebs projecting from the outer surface of the central protoplasmic membrane. In Figure 6, the completed central body with abundant PHB granules has become clear. The outer wall of the central body shows numerous tiny blebs. Surrounding the central body is also seen the development of a capsular polysaccharide layers which appear to be differentiating into two layers as indicated by arrows. Figure 7 shows, three different complete cysts with the development of a hard coat surrounding a central body. Each cyst is also seen engulfed in or surrounded by an extra-exinic granular material which appears brown in color when viewed under a phase contrast microscope, Figure 8 is a higher magnification of a single, fully matured cyst, in which a hard cyst coat or exine is clear and is separated from the central body by a highly refractile less electron dense intine-like region. These cysts when stained according to the procedure of Vela and Wyss (19), showed the central body to be green and the outer exinic coat, red or brown while the intermediate intine region remained unstained.

Figures 9, 10 and 11 show the different stages of the germinating cysts. As reported for Azotobacter spp. (ref 1, 2), the formation of giant cysts or cysts with multiple central bodies are also evident. The giant cysts of A. brasilense may form from one to four central bodies per cyst. During the breaking apart of the cyst wall, sheets of laminated forms were occassionally observed (Figure 12). Similar observation have also been reported for Azotobacter spp. (11).

Separation of the different cell forms by sucrose density gradient yielded three different forms differing in their density. The brown cysts along with the aggregated granular material pelleted at the bottom of the 70% sucrose gradient (Figure 13). Mixtures of vegetative cells and few isolated granular particles were observed in the 65% gradient while the band at the 55% gradient showed mostly cell forms which were coccoid-like bodies and empty sac like structures (Figure 14).

In liquid culture medium, the formation of pigmentation and complete encysted cells were not observed in either A. brasilense or in A. lipoferum instead the inoculum cells flocculated due to the production of extracellular polysaccharides (12). The cells in the flocs were desiccation resistant and were viable even after 10 months of storage. As reported by us (12), the cells in the flocs develop enormous amount of PHB but do not complete its encystation. There is no development of exine or intine, instead the polysaccharide surrounds the cells and they did not show brown pigmentation.

Characterization of the brown pigment

The brown granular particles were soluble in hot 0.5 N KOH or NaOH. The absorption spectrum of this crude preparation of the pigment showed a peak between 310 and 320 nm (Figure 15). Similar absorption spectrum has been reported for a melanin producing mutant of Vibrio cholerae (5). The pigment also showed positive reaction to other properties of melanins like their insolubility in water or any organic solvents, bleaching property in chlorox and decolorization in oxydizing agent like 10% H_2O_2. Hence, the brown pigment produced by A. brasilense is speculated at this stage to be melanins.

Discussion

In the present paper, we have reported most of the morphological changes that take place during encystation and germination in A. brasilense. These changes appear to be similar in some respects to the changes associated with the encystation in Azotobacter vinelandii (13). The availability of fructose as a direct carbon source readily induces cyst formation in

A. brasilense. In case of A. vinelandii, using butanol as a sole carbon source, 90 to 95% of the vegetative cells could be induced to form cysts (13, 14, 15), whereas in A. brasilense butanol fails to induce cyst formation and in a medium containing fructose as a carbon source, only a fraction of the cells undergo encystation. The cysts were always observed mixed with vegetative cells even after a prolonged period of incubation of more than 2 weeks, as revealed microscopically and by sucrose density gradient separation. The apparent inability of azospirilla to show 90 to 100% of the cells encysted at any one time may be related to superimposed cycles of encystation and germination by different cells within a population as reported earlier by Papen and Werner (9).

In addition to the basic morphologic similarities of cysts between Azospirrillum spp. and Azotobacter spp., a unique variation in A. brasilense is observed, the formation of melanin granules surrounding the cysts and excreted into the medium. This phenomenon is not only a new observation in A. brasilense but it is not very common in bacteria, in general. The production of melanin during encystation may have a function of protection against deleterious conditions like light, heat, microbial degradation etc. (5). Furthermore, the production of melanin pigments seem to be strain or species dependent since A. lipoferum did not show any visible pigmentation either in pellicle or in colonies.

Acknowledgements

We thank Laura Wood, Nelson Biological Laboratories, Rutgers University for taking the EM pictures; Barbara Zilinskas and her colleagues, Department of Biochemistry and Microbiology, Rutgers University, for help in the pigmentation studies and Bonnie Boettger, for typing of the manuscript. We also thank N.R. Kreig, Virginia Institute, Blacksburg, Virginia for sending the bacterial strains.

This work is published under the New Jersey Agricultural Experiment Station, Publication No. 0-01204-02-85, supported by state funds and United States Hatch Act funds. We acknowledge the financial support to L.S. as a visiting scientist and Post Doctoral Fellow by the NJES, Department of Biochemistry and Microbiology and the International Agriculture and food programs at Cook College, Rutgers University.

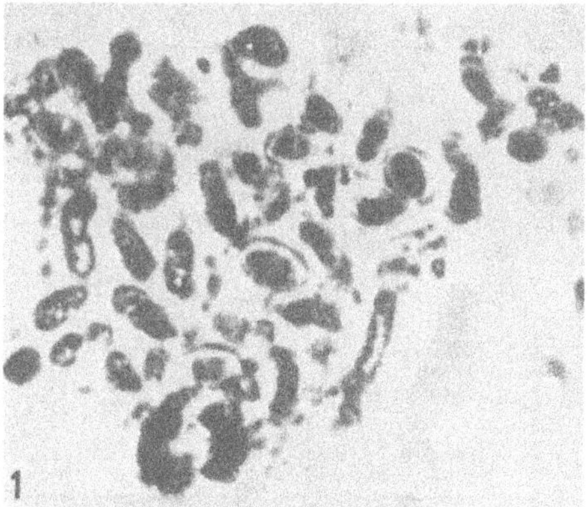

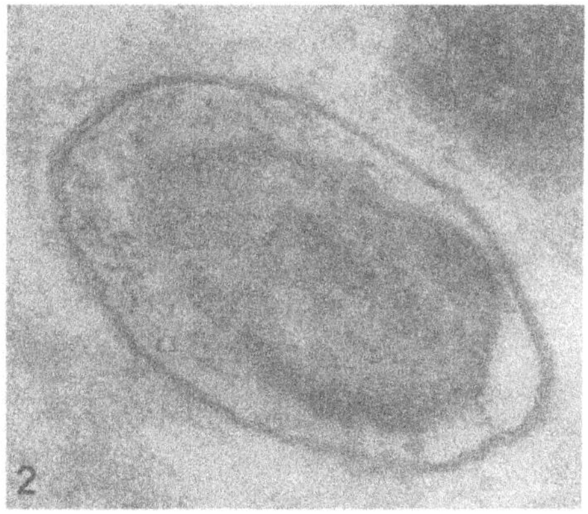

Figure 1. Phase contrast microscopic observation of a desiccated pellicle of *A. lipoferum* (X1,000).

Figure 2. Ultrathin section of a desiccated pellicle of *A. lipoferum*, showing cells with no PHB accumulation but shrinking of the protoplasm into a central body. This is surrounded by an outer coat-like layer (X14,000).

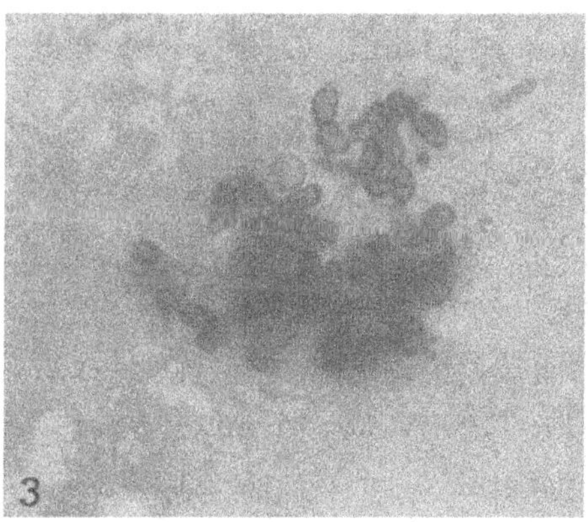

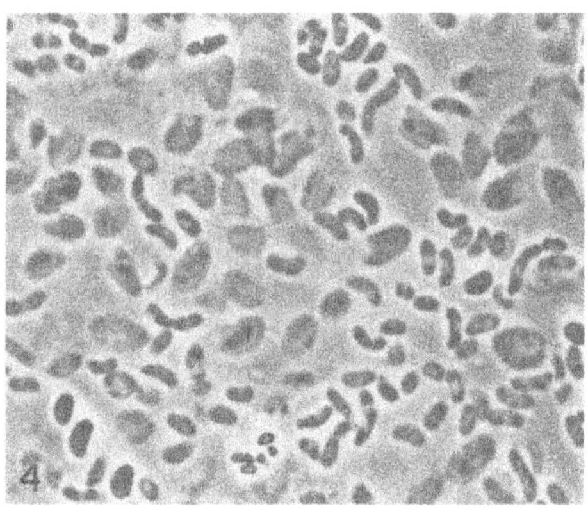

Figure 3. Phase contrast microscopic photograph of the encysted cells of *A. brasilense* from a brown desiccated colony (X1,000).

Figure 4. Phase contrast microscopic observation of the germinating cysts of *A. brasilense* from a brown colony. Note the vegetative cells emerging out of a horseshoe-shaped sac-like exinic layer (X1,000).

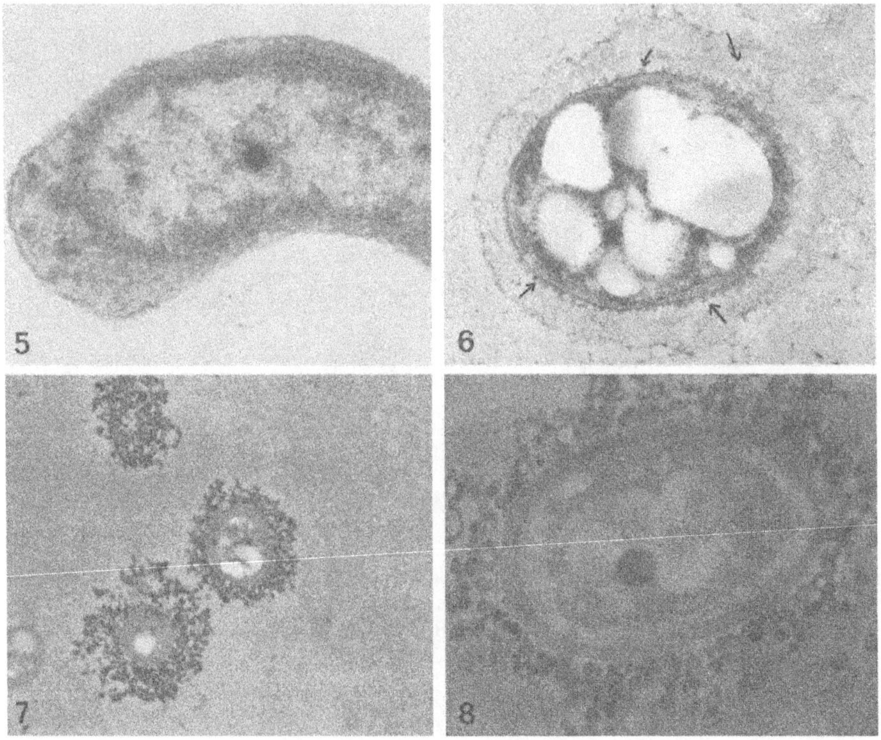

Figure 5. Ultrathin section of an encysting *A. brasilense* cell on a solid agar medium. Note the lipoprotein-like blebs from the shrinking protoplasmic membrane (X30,000).

Figure 6. Thin section of the encysting *A. brasilense* cell with the formation of a complete central body with PHB granules and numerous tiny blebs (arrows). The capsular polysaccharide layer is evident (X24,000).

Figure 7. Thin section of the mature cysts of *A. brasilense* showing the melanin-like granules surrounding all the three cysts (X5,000).

Figure 8. Thin section of a mature cyst of *A. brasilense* at a higher magnification (X20,000). The central body with PHB granules and an electron dense body is separated from the hard outer coat (exine) by a less electron dense intine like region.

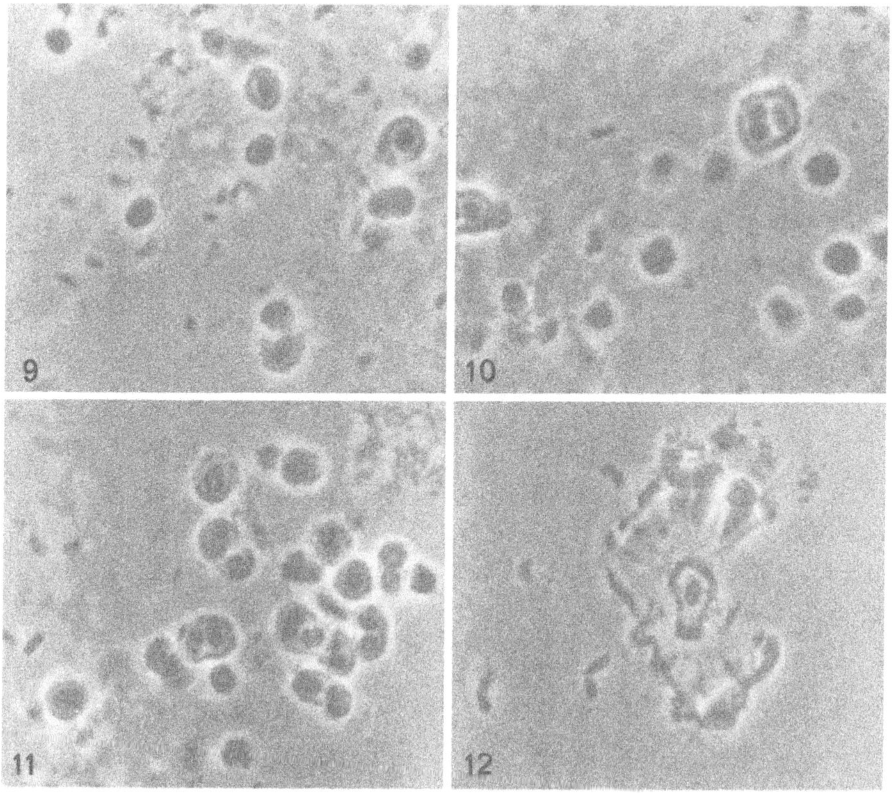

Figure 9. Phase contrast microscopic observation of the germinating cysts obtained from brown pellicle of _A. brasilense_. The central dense body is seen surrounded by a coat (X1,000).

Figure 10. Germinating _A. brasilense_ cysts with the formation of two or three central bodies (X1,000).

Figure 11. Phase contrast microscopic observation of the _A. brasilense_ giant cysts with multiple central bodies (X1,000).

Figure 12. Germinating _A. brasilense_ cysts showing the laminated sheets of the exine layer (X1,000).

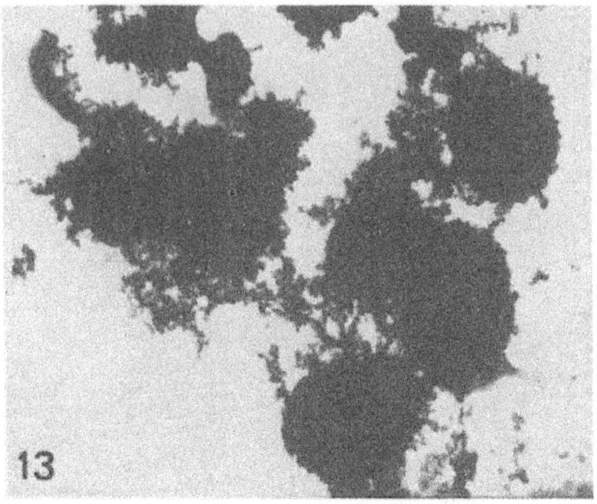

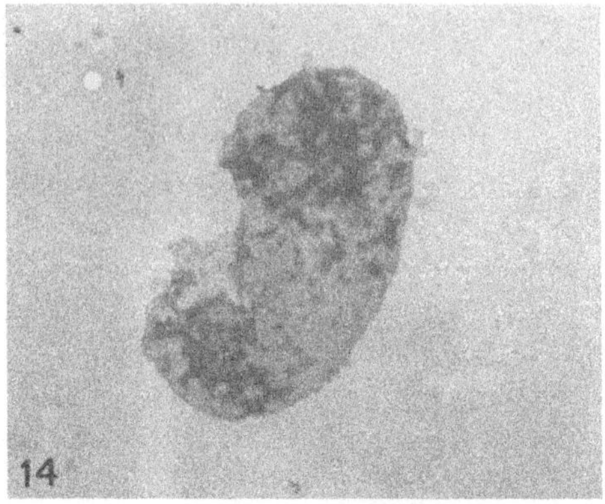

Figure 13. Negative staining with uranyl acetate of the dense, brown cysts and the granular material from the 70% sucrose pellet (X15,000).

Figure 14. Negative staining with uranyl acetate of the empty sac-like structure from the 55% sucrose gradient (X20,000).

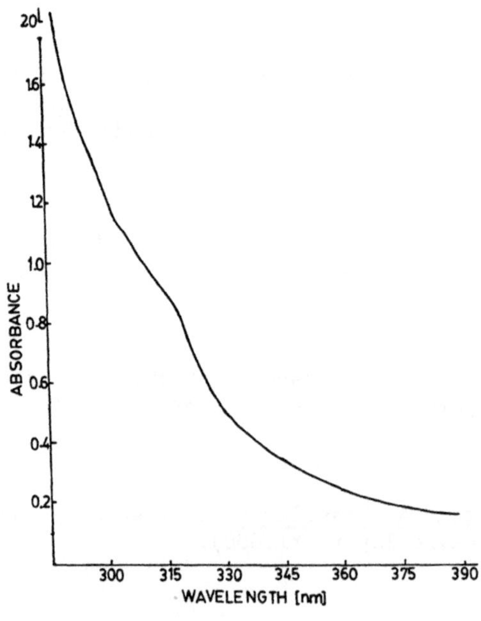

Figure 15. Absorption spectrum of the alkali fraction (Hot 0.5 N KOH) of the brown pigment from A. brasilense.

References

1. Beaman, B.L., Jackson, L.E. and Shankel, D.M. (1968). J. Bacteriol. 96:266-269.

2. Cagle, G.D. and Vela, G.R. (1971). J. Bacteriol. 107:315-319.

3. Cole, R.M. and Popkin, T.J. (1981). Electron Microscopy p. 34-51. In p. Gerhardt, R.G., E. Murray, R.N. Costilow, E.W. Nester, W.A. Wood, N.R. Kreig and G.B. Phillips (ed.), Manual of methods for general bacteriology. American Society for Microbiology, Washington, D.C.

4. Hitchins, V.M. and Sadoff, H.L. (1970). J. Bacteriol. 104:492-498.

5. Ivins, B.E. and Holmes, R.K. (1980). Infect. Immun. 27:721-729.

6. Lamm, R.B. and Neyra, C.A. (1981). Can. J. Microbiol. 27:1320-1325.

7. Layne, J.S. and Johnson, E.J. (1964). J. Bacteriol. 87:684-689.

8. Neyra, C.A. and Van Berkum, P. (1977). Can. J. Microbiol. 23:306-310.

9. Papen, M. and Werner, D. (1982). Arch. Microbiol. 132:57-61.

10. Parker, L.T. and Socolofsky, M.D. (1966). J. Bacteriol. 91:297-303.

11. Pope, L.M. and Wyss, O. (1970). J. Bacteriol. 102:234-239.

12. Sadasivan, L. and Neyra, C.A. (1985). J. Bacteriol. 163 (in press).

13. Sadoff, H.L. (1975). Bacteriol. Rev. 39:516-539.

14. Socolofsky, M.D. and Wyss, O. (1961). J. Bacteriol. 81:946-954.

15. Socolofsky, M.D. and Wyss, O. (1962). J. Bacteriol. 84:119-124.

16. Stevenson, L.H. and Socolofski, M.D. (1966). J. Bacteriol. 91:304-310

17. Vela, G.R. (1974). Appl. Microbiol. 28:77-79.

18. Vela, G.R. and Cagle, G.D. (1972). Can. J. Microbiol. 18:371-373.

19. Vela, G.R. and Wyss, O. (1963). J. Bacteriol. 87:476-478.

PLEOMORPHISM IN AZOSPIRILLUM

J.H. Becking
ITAL
Research Institute of the Ministry of Agriculture
Division for Agricultural Research
P.O. Box 48, 6700 AA WAGENINGEN, The Netherlands

Summary

Pleomorphism in azospirilla appears to be induced by medium composition rather than pH. Complex organic media stimulate it and it is more pronounced in *Azospirillum lipoferum* than in *A. brasilense*. Besides 'capsules', protoplasmic globular structures (free protoplasts) sometimes may be produced. Under certain conditions the smaller cells appear to have a flexible cell wall, whereas the helical, spiril-like cells are rigid and show a cork-screw type of motility in straight lines - often reversing direction - as demonstrated in a motion picture. Cell volume and growth rate seem to be correlated, the larger the pleomorphic cells the longer the generation time. Differences in cell morphology suggest that some strains now denoted as *Azospirillum* might show relationships to other groups. The taxonomic position of azospirilla in context to other taxa is discussed.

Introduction

When Beijerinck (7) isolated the so-called *Spirillum lipoferum* in 1925 (previously named *Azotobacter spirillum* (6) in 1922), he could not imagine that his *Spirillum*, now called *Azospirillum*, would become so world famous that workshops or even symposia are held dealing with it. Although Beijerinck (7) doubted that his *Spirillum* could fix dinitrogen, it was merely 40 years later (in 1963) that it was incontestably proved to be a nitrogen

Key words: Pleomorphism Capsules Free-Protoplasts Motility Cell-wall flexibility Linkage cell volume/generation time Taxonomy.

Azospirillum III: Genetics, Physiology, Ecology
Ed. by W. Klingmüller

fixer by obtaining a pure culture, which could utilize $^{15}N_2$ (1,2). It was, however, compared with some other diazotrophs a minor nitrogen fixer, which required growth substances for growth in N-free medium. A new area started with the observation of Döbereiner in 1973 (see ref. 10), that this organism was primary associated with plant roots, particularly Gramineae in tropical countries and by its action stimulated the growth of these plants.

In Beijerinck's (7) original description already two prominent characteristics of the organism were mentioned. First, its great preference for organic acids (especially malic acid) as sole carbon source for heterotrophic growth, and secondly, its extreme pleomorphism. The latter was very well illustrated by some drawings showing slightly curved vibrioid cells with internal reserve granules, straight rods and more helical spiril-like forms. In 1978 Tarrand *et al.* (22) proposed in a taxonomic study the name *Azospirillum* for this species and recognized on basis of different characteristics two species: *Azospirillum lipoferum* (probably identical with Beijerinck's original strain) and *A. brasilense*. Later, in 1983, a third species *A. amazonense* (10) have been described.

The purpose of the present study was to examine the pleomorphism of some well-known *Azospirillum* strains on various media. Further, to evaluate the so-called pH effect of the medium on pleomorphism and to obtain insight in the observed coccoid or ovoid forms regularly encountered in cultures of *Azospirillum*. The latter forms are defined by some authors to be cysts, but by others to be capsules. Finally, the taxonomic position of this genus and its supposed phylogenetic affinities are discussed.

Materials and Methods

Strains: Two species of *Azospirillum* were examined, the third species, *A. amazonense*, was not studied as it is reported to be non-pleomorphic (10).

Azospirillum lipoferum: Sp. 41 (soil, The Netherlands, J.H. Becking), Sp. B7c (maize, France, T. Heulin), Sp. 4B (rice, France, T. Heulin), Sp 4T (mutant from 4B, see ref. (14)), Sp. 20/18 (maize, Germany, G. Jagnow), Sp. 59b (ATCC 2970) (wheat, Brazil, J. Döbereiner), Sp.Br.17 (ATCC 29709) (maize, Brazil, J. Döbereiner).

A. brasilense: Sp. KUL 2 (wheat, Belgium, L. Reynders), Sp. KUL 9 (soil, Greece, L. Reynders), Sp. Cd (ATCC 29710) (*Cynodon*, U.S.A., D.L. Eskew),

Sp.7 (ATCC 29145) (Digitaria, Brazil, J. Döbereiner), Sp.M 81 (maize, Brazil, J. Döbereiner).

Media: The following media (solid 12.0 g agar/litre; semi-solid 1.75 g agar/litre) were employed:

Nitrogen-free media: Modified Heulin medium (13) using Na-malate 1.5 g/litre and Nfb-medium (16) using Na-malate 5.0 or glucose 10.0 g/litre, all pH = 6.8.

Complex media (g/litre): (a) Peptone (Oxoid) 10, NaCl 5.0; (b) Beef-extract, Lab-Lemco (Difco) 1.0, Yeast extract (Difco) 2.0, NaCl 5.0; (c) Yeast-extract-glucose, Yeast extract (Difco) 7.0, $MgSO_4 \cdot 7H_2O$ 0.2, K_2HPO_4 0.5, biotin 0.0001, pyridoxin 0.0002, Na-malate 5.0; (d) Plate-count-agar (PCA), yeast extract 2.5, tryptone 5.0, glucose 1.0. All pH 6.8 - 7.0.

Morphological observation were made using a Wild or a Zeiss microscope with phase-contrast optics. Cell dimensions were determined by photographing a stage micrometer and printing it at the same degree of enlargement as the other photographs.

Cell counts were made with a Bürker-Türk counting chamber of unstained cells appropriately diluted. Growth was followed by cell counts and by measuring the extinction (OD_{660}) with a Vitatron Digital Concentration Photometer (Vital Scientific Benelux B.V., Dieren, The Netherlands).

Morphological observations

In view of space limitations only some particular strains will be discussed in detail.

Azospirillum lipoferum Sp. 41

This strain is our own isolate from Dutch soil (1,2). Since it requires biotin (or yeast extract) for growth in N-free medium and it utilizes glucose as sole C-source, it can be allocated as Azospirillum lipoferum. On N-free mineral glucose medium (supplemented with yeast extract) the cells appear as plumb, slightly curved rods with usually rounded ends. The cells are c. 1.0 µm in diameter and 2.1 - 3.5 µm in length, however some cells may attain a length of 3.9 - 4.7 µm (Fig. 1, A). Each cell contains 3-4 poly-β-hydroxybutyrate globules (see for chemical analyses ref.(4)). In the shorter or longer cells 2 or 5 PHB-granules may be present, respectively. On beef-extract-agar the cells are more spiril-like with pointed ends, but

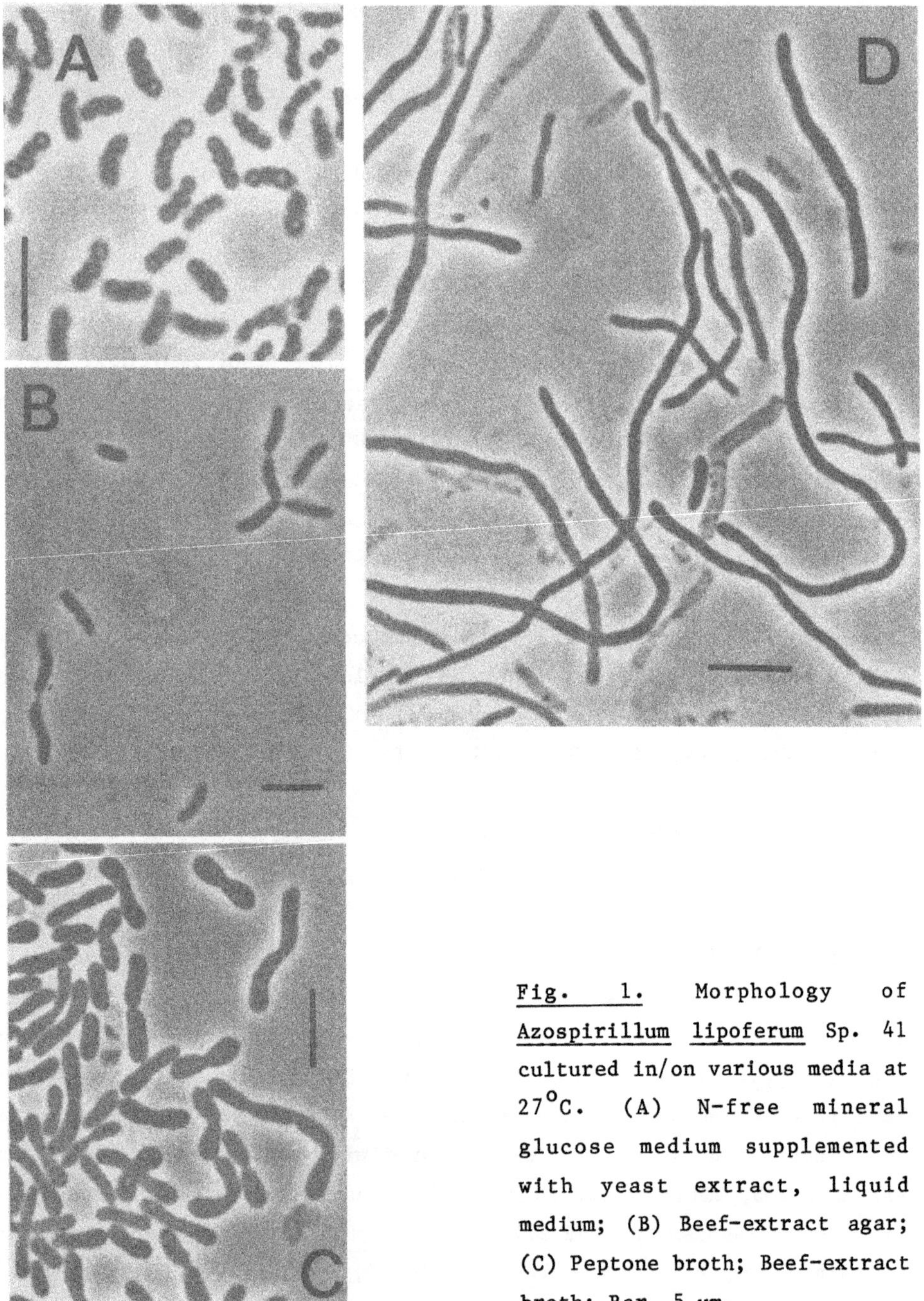

Fig. 1. Morphology of Azospirillum lipoferum Sp. 41 cultured in/on various media at 27°C. (A) N-free mineral glucose medium supplemented with yeast extract, liquid medium; (B) Beef-extract agar; (C) Peptone broth; Beef-extract broth; Bar, 5 µm.

usually only possessing half of a complete turn. In this medium the cells often occur in characteristic angular formations, so-called 'star'-formation (V, Y configurations), which probably arise by non-separation of cells after fission or by 'snapping' post-fission movements (Fig. 1, B). The cells are about 3.6 - 5.0 µm in length and 0.9 - 1.0 µm in diameter and lack completely phase-refractile particles. On peptone-based agar the cells appear to be shorter, 2.5 - 3.5 µm in length, but with about the same width of 1.0 - 1.1 µm. Again, there is some tendency to V or 'snapping' configurations. In beef-extract broth, especially in the older cultures, this strain produces very elongated cells of normal diameter, often 90-110 µm in length, apparently by failure of separation of these cells after cell division. These long 'threads' may be constituted of 30-40 individual cells (Fig. 1, C). These cultures show also a high proportion of lysed cells as evident by the absence of cell contents.

Azospirillum lipoferum Sp. B7c and Sp. 4B (4T)

Both strains were obtained from the Laboratoire d'Ecologie Microbienne, C.N.R.S., Vandoeuvre-Cédex, Nancy (France) by the courtesy of Dr. J. Balandreau and Dr. T. Heulin. The origin of the strains is discussed in ref. (13).

Strain Sp. B7c showed on a number of media such as Plate-count-agar (PCA) and Yeast-extract-glucose agar an identical and rather uniform morphology. The cells are round ovoid or coccoid in shape, sometimes with a tendency to duplococcus formation (Fig. 2, A). The ovoid/coccoid forms are 1.2 - 1.6 µm in diameter and the 'twin'-forms are 1.2 - 1.4 µm in diameter and 2.2 - 2.7 µm in length. Regularly, much larger spherical forms are observed (see arrow, Fig. 2, A) which might be capsules or cyst-like structures. In contrast to the above-mentioned observations, strain Sp. 4B (4T) on the same media showed a morphology of long, slightly curved rods c. 0.9 - 1.0 µm in diameter and varying in length between 3.0 and 5.0 µm (Fig. 2, B). In older cultures, often already within a week, rather numerous spherical bodies are developed. These bodies are c. 2.0 - 4.0 µm in diameter and may be capsules or cysts (see later).

Azospirillum lipoferum Sp. 59b

This strain (ATCC 29709) was isolated by Dr. J. Döbereiner from wheat roots in Brazil. It is the neotype on which the type description of A. lipoferum is based (16, 22). A subculture of this strain was obtained

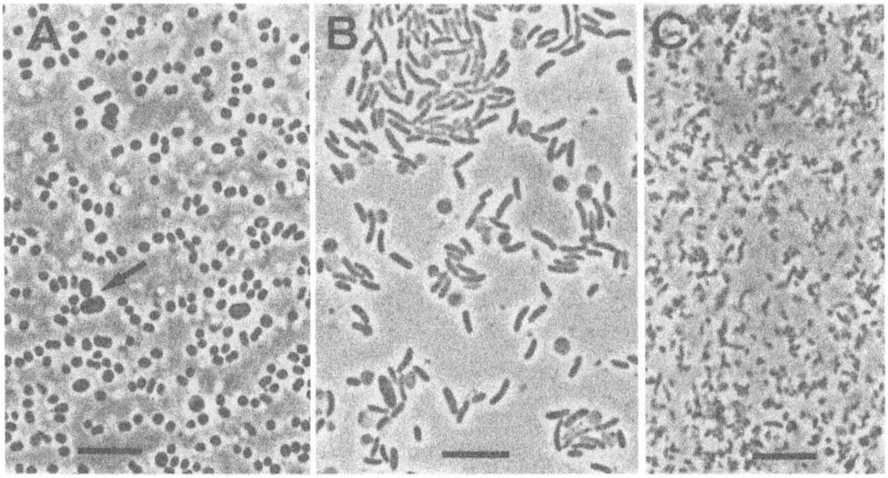

Fig. 2. Azospirillum lipoferum : (A) Strain B7c, (B) Strain 4B, and (C) Strain 59b; all cultured on Plate-count-agar at pH = 6.2, 27°C. Strain Sp. 4B shows the development of globular structures, probably capsules. Bar, 10 µm.

through the courtesy of Prof. Dr. G. Jagnow, FAL, Braunschweig, Germany. This subculture showed rather small, very motile cells, c. 0.6 - 0.8 µm in diameter and 1.9 - 2.3 µm in length. These cells are smaller than most of the other Azospirillum strains examined by me. The cells are rod-shaped, slightly bend or vibrioid (Fig. 2, C). The strain showed little morphological variation in cell type on a wide spectrum of complex media. It remains to be investigated whether this subculture is (still) identical with the original strain isolated by Dr. J. Döbereiner.

Azospirillum brasilense Sp. KUL 2 and Sp. KUL 9

Both strains were isolated by Dr. L. Reynders of the Catholic University of Leuven, Faculty of Agriculture, Belgium. The strains manifest themselves on a wide range of media to be more uniform and regular in appearance than the majority of Azospirillum lipoferum examined by me. Cells of strain Sp. KUL 2 are straight or slightly curved rods c. 0.9 µm in diameter and 2.9 - 3.3 µm in length (Fig. 3, A). This shape undergoes little variation in complex media, although regularly spherical or coccoid forms (c. 1.6 - 1.9 µm in diameter) are observed (see arrow, Fig. 3, A) resembling capsules or

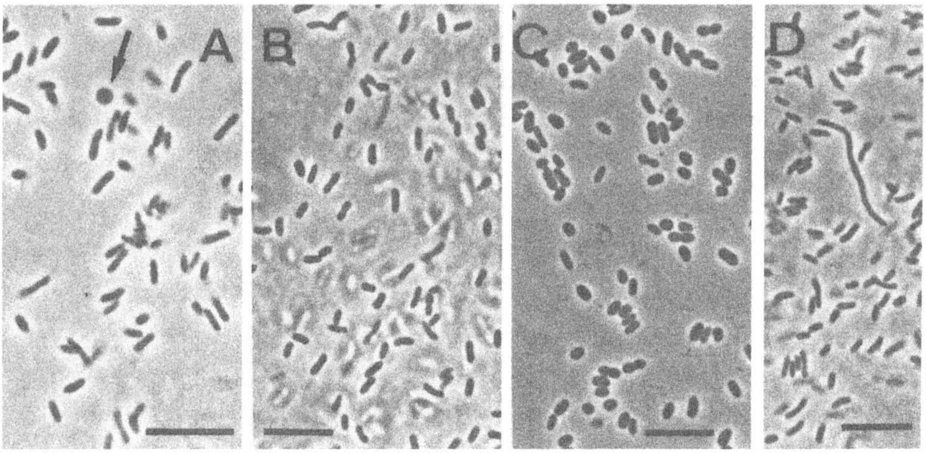

Fig. 3. Azospirillum brasilense : Strain KUL 2 on Plate-count-agar, capsule (arrow); (B) Strain KUL 9 on peptone agar; (C) Strain KUL 9 on beef-extract agar; (D) Strain KUL 9 on yeast-extract glucose agar. Cultured at pH = 6.2, 27°C. Bar, 10 µm.

cysts. Strain Sp. KUL 9 gave also the same regular pattern of slightly curved rods of 2.3 - 3.3 x 0.8 µm in size. On peptone-based media the cells have approximately the same length, but appear tobe somewhat wider in diameter, i.e. 1.0 - 1.1 µm. Larger differences occur on plate-count-agar, but particularly on beef-extract agar the cells were much wider in diameter, i.e. 1.4 - 1.5 µm, whereas cell length remained the same (compare Fig. 3, B/C). In spite of the more uniform cell-shape of the Sp. KUL strains, occasionally very large elongated cells of a length of 23-30 µm may occur (Fig. 3, D). The latter structures probably consist of cells which have divided 7-10 times without separation.

Azospirillum brasilense Sp.7, Sp.Cd, and Sp.M 81.

All these strains have been used as reference strains for the type description of A. brasilense. (16, 22). Compared to the strains Sp. KUL 2 and Sp. KUL 9, these strains are more pleomorphic. In complex media the rods become swollen and in older cultures, particularly on PCA, the rods readily develop into globular, capsule-like structures. In strain Sp.7 coccoid formation under such conditions is very dominant, while in strain Sp.Cd often a more equal mixture of rods and coccoid structures finally occur. In view of its more uniform cell morphology strain Sp.M 81 resembles the Sp. KUL strains. Usually its cell size is 0.8 - 0.9 x 2.0 - 3.2 µm, however, sporadically very elongated cells of 52 - 55 µm may occur (see Fig. 4).

Effect of pH on cell morphology

According to Krieg and Döbereiner (16) alkalization of the malate medium due to oxidation of malate, may relate to the development of pleomorphism in *Azospirillum lipoferum*. Because this phenomenon failed to occur in semi-solid N-free glucose medium, which does not become alkaline. Two figures (Fig. 2.10 & 2.11) published by these authors concerning *A. lipoferum* ATCC 29707 indicate, however, that cells cultured on N-free glucose medium (Nfb-medium) are somewhat shorter and broader than those grown on peptone-succinic acid medium (MPSS) under the same conditions. Hence, apart of a pH effect a complex medium effect on cell morphology may exist. We studied the cell morphology of the two *A. brasilense* Sp. KUL strains on a complex medium (yeast-extract-glucose) with varying pH values

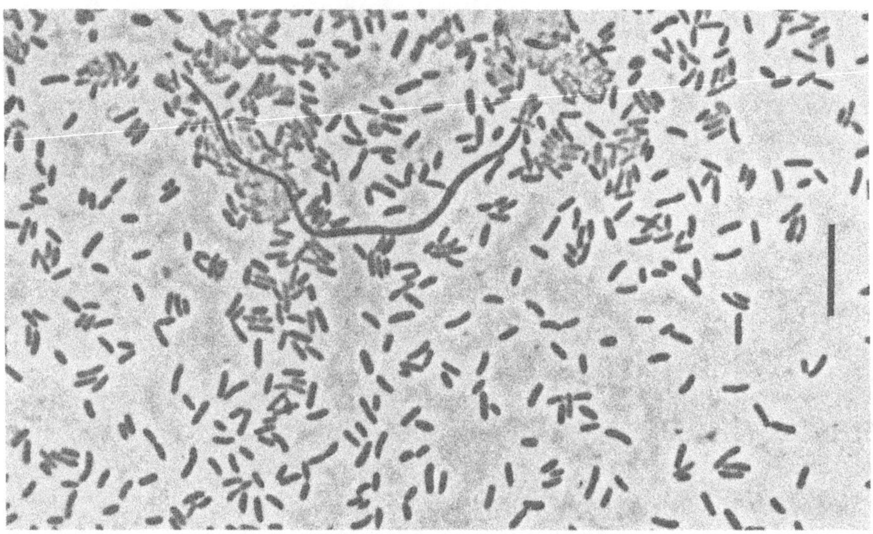

Fig. 4. *Azospirillum brasilense* : Strain Sp. M 81 cultured on yeast-extract glucose agar, pH = 6.2, 27°C. Note the very elongated cell, probably fission products which have not separated. Bar, 10 µm.

(pH = 5.1, 6.2, 8.0, 9.0 and 10.0). None of the strains grew on agar of pH 5.1. Wider tests with more azospirilla strains and pH values tested, demonstrated that the lower limit of growth for *Azospirillum* is c. pH = 5.6. Comparison of cell morphology of 3-days-old cultures of both KUL strains grown at 29°C on agar of pH = 6.2 and 10.0 (the extremes) showed no differences in cell size and morphology. At both pH values the cells measure 0.8 - 0.9 x 2.0 - 3.5 μm.

Capsulated, cyst-like or ovoid/coccoid forms

Non-motile, enlarged, pleomorphic forms (C-forms) are reported in older cultures of *A. brasilense* (11, 22) or on the surface of N-free agar media in association with plant callus (8). Lamm and Neyra (17) mentioned that cultures of both *Azospirillum* species enriched in C-forms or cysts exhibit a resistance to desiccation and temperature not found in cultures lacking these structures. Krieg and Döbereiner (16) suggest that these C-forms may be related to Gram-decolorization and Gram-variability, which is also more pronounced in *A. brasilense* than in *A. lipoferum* and that these forms are rather encapsulated cells than real cysts on basis of ultra-microscopic studies (see also ref. (8)). Our observations suggest that there exist two types of globular structures. The first is a very thin-walled structure. It may be a naked bacterium protoplast, which has obtained a globular form by surface tension. These forms are often found in *A. lipoferum* cultures, where a large number of cells has undergone lysis (Fig. 5). Although these structures tend to appear more frequently in *A. lipoferum* cultures, they may also occur in *A. brasilense*. The second structure (see Fig. 2, B and 3, A) is a thick-walled, cyst-like coccoid structure similar as those described by other authors (8, 11, 16, 17). These structures show great resemblance to structures found in *Azotobacter* (3) and we are inclined to consider them capsules. Lamm and Neyra (17) consider them cysts, but also in *Azotobacter* it is difficult to decide on morphological appearance only, between cysts and capsules, i.e. whether they are mono- or multicellular. Although electron microscopical studies have shown that some of the structures are real capsules enclosing a number of cells (16), the presence of cysts can not be ruled out. Also in *Azotobacter* both structures occur side by side. Further ultra-structural studies are needed for the identification of cysts. Cyst structure should comprize the presence of a cyst coat and a centre core (DNA) surrounded by layers of an exocytorium and exine.

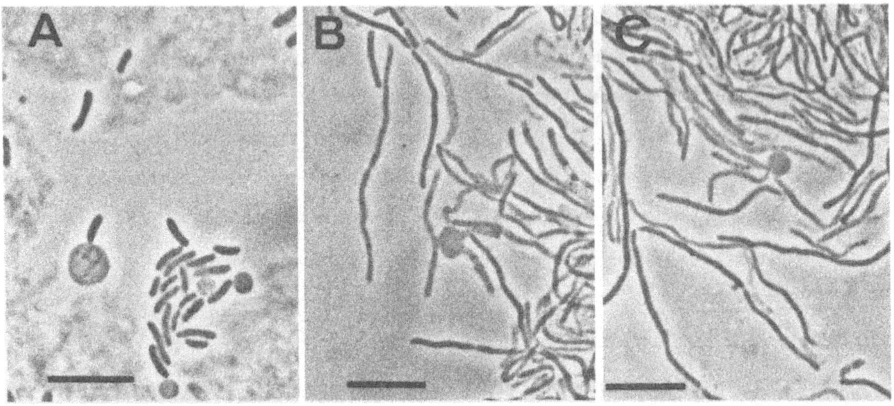

Fig. 5. Globular protoplasmic structures, probably free protoplasts in two Azospirillum lipoferum strains. (A) Strain 4B cultured in yeast-extract glucose agar ; (B) and (C) Strain Sp. 41 in N-free mineral glucose liquid medium supplemented with yeast extract. Bar, 10 µm.

Motility and cell-wall flexibility

Phase-contrast light microscopy and dark-field illumination revealed a characteristic type of movement in azospirilla. The very motile cells swim in straight lines by a single flagellum (3), but at the same time they perform rotary movements in which the cell wall appears to be flexible. These movements have some resemblance with the coordinated flexible movements of a fish corpus during swimming. Such alterations are difficult to capture in static pictures and for this reason a 16-mm cinematographic motion picture was produced with 24 frames per second presenting the movements of small swimming cells of A. lipoferum Sp.41 grown in N-free glucose broth as observed by phase-contrast and dark-field light microscopy. This cinema-film (movie) was demonstrated during the session and some of the consecutive frames showing the sequence of movements are printed as negative print of the positive picture in Fig. 6. These motion-picture prints give some impression of the flexibility of the cell wall, which in 'normal' bacteria seems to be a rather rigid structure. This cell-wall flexibility is certainly not due to special structures in the cell wall such as the presence of contractive fibrils, etc. It may be caused by the particular nature of the peptidoglycan molecules constituting these cell walls.

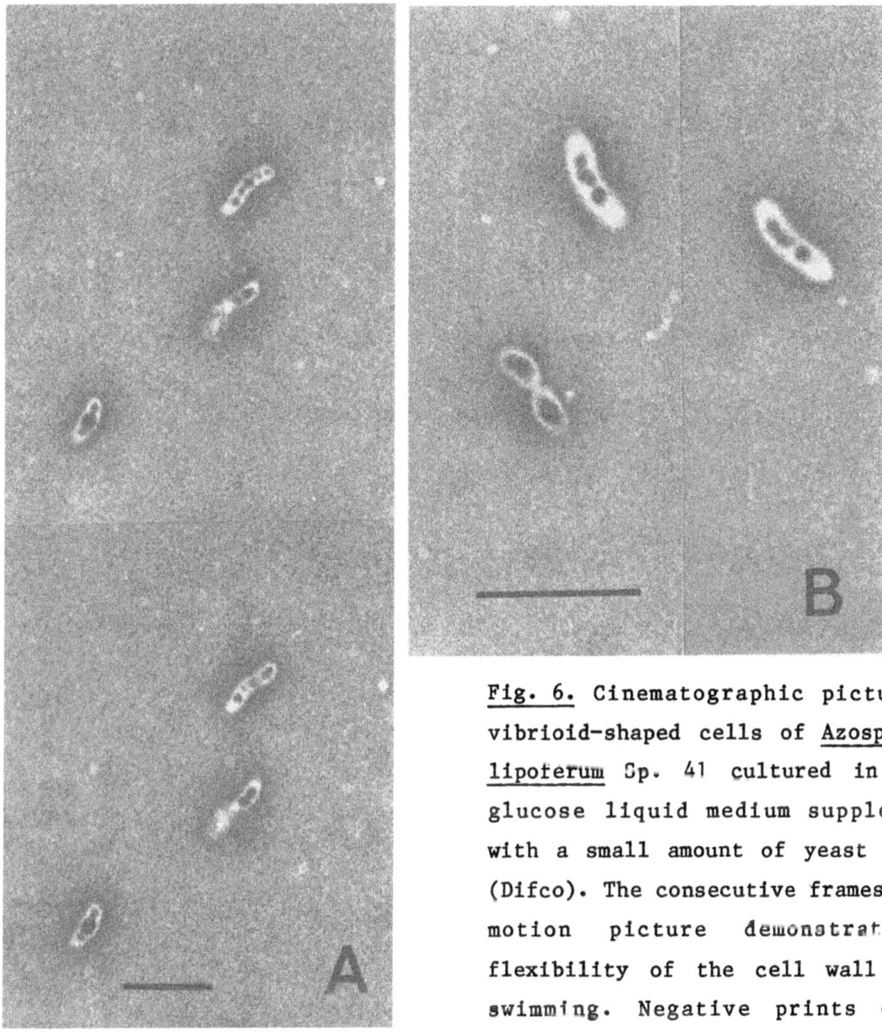

Fig. 6. Cinematographic pictures of vibrioid-shaped cells of Azospirillum lipoferum Sp. 41 cultured in N-free glucose liquid medium supplemented with a small amount of yeast extract (Difco). The consecutive frames of the motion picture demonstrate the flexibility of the cell wall during swimming. Negative prints of the positive picture. In (B) a division of a vibrioid cell is shown. Phase-contrast microscopy; Bar, 5 µm.

It has been suggested that the elongated, pleomorphic forms of Azospirillum are non-motile (16). The cinematographic shots presented in the same motion picture show that the spiril-like forms of Azospirillum having a complete wavelength with 3-4 turns, exibit highly motile movements very typical for Spirillum species. In this form the cell wall is rigid and the cells perform a cork-screw type of rotation in straight lines frequently reversing their direction of orientation (Fig. 7). These helical cells have no true anterior cell pole as evident from the very active and strong back and forwards movements demonstrated in the motion picture.

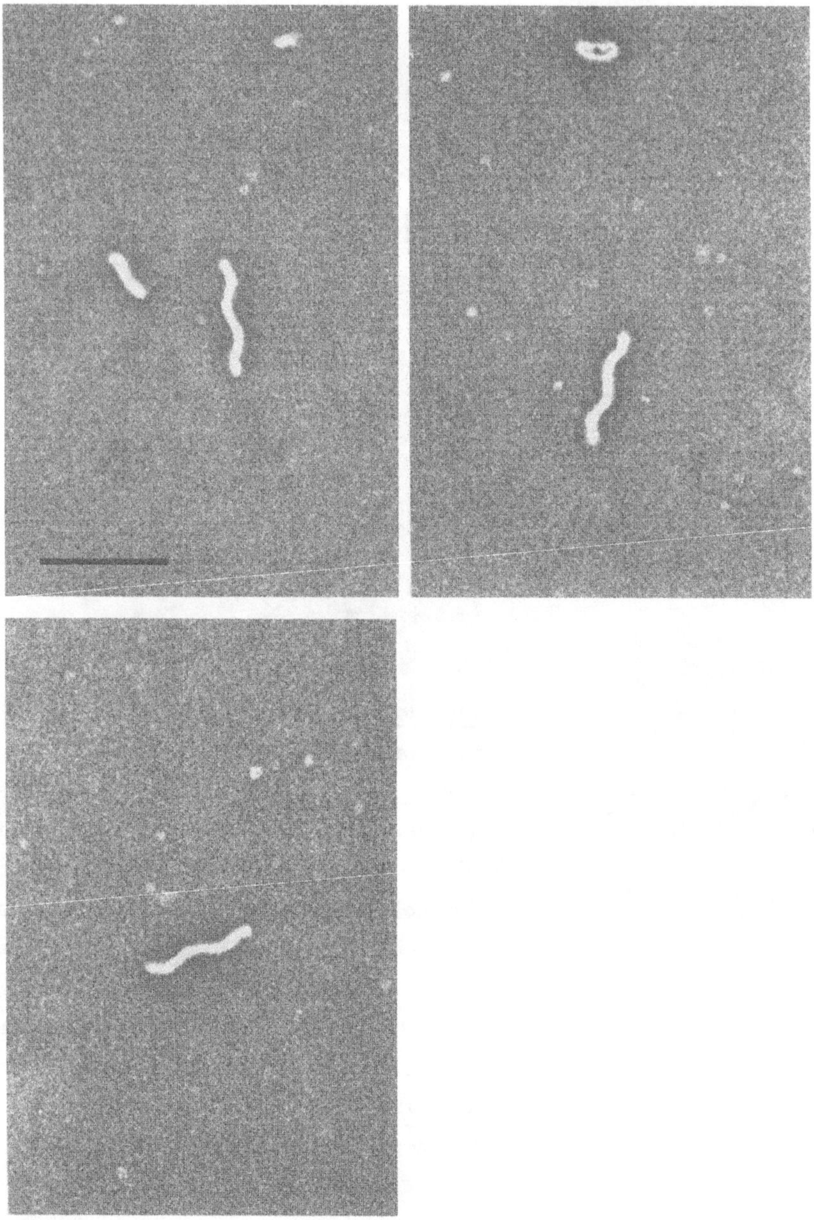

Fig. 7

Cinematographic pictures of Azospirillum lipoferum Sp. 41 cultured in a N-free mineral glucose liquid medium supplemented with a small amount of yeast extract (Difco). The three photomicrographs are the subsequent frames of a motion picture printed as negative print of the positive picture. The helical cells are extremely motile performing a cork-screw type of rotation, while swimming rapidly back and forward in straight lines. Phase-contrast microscopy, Bar = 10 µm.

Cell size and growth characteristics

An increase of cell size must have some repercussions on the rate of protein synthesis by the cell. Moreover, an increase of protein synthese must go hand in hand with an increase of RNA content of the cell. Therefore, pleomorphism must affect in some way the growth rate of the organism. Growth rate determination were performed applying cell counts and extinction measurements (see Materials & Methods). Moreover, the cell volume measurements of cells grown on different media were performed using a C-1000 channelyzer linked to a ZBI Coulter counter (Coulter Electronics, Harpenden, Herts., England) after calibration of the latter with latex beads of 1.87 µm diameter. After selecting an appropriate window width, treshold, aperture, and attenuation c. 20,000 - 50,000 cells per sample were scored to give an average modal cell size.

The modal cell volumes of A. lipoferum Sp.59b and A. brasilense Sp.KUL 9 on Plate-count-agar (PCA), pH= 6.8 as registered by the Coulter Counter Channelyzer are presented in Fig. 8. In Table 1 the modal cell volumes (using the equation TF = MCV/MCC, in which TF = a conversion factor, MCV =

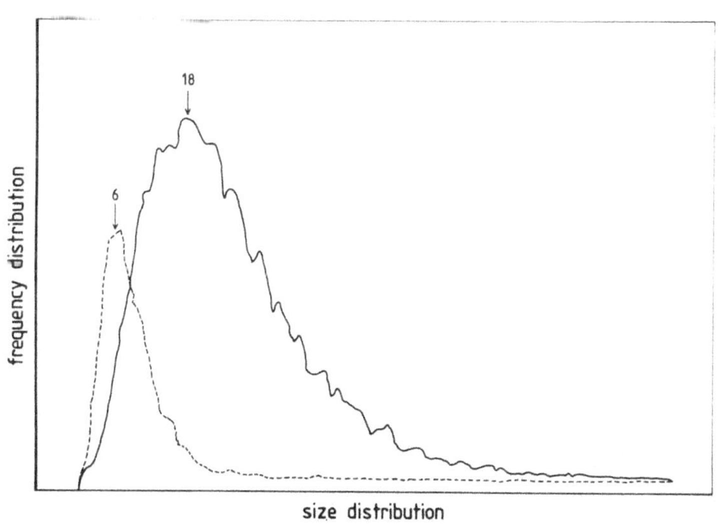

Fig. 8. Frequency distribution curves of relative number of cells in various cell-volume ranges as visualized by the Coulter Counter Channelyzer. Broken line Azospirillum lipoferum Sp.59b; solid line A. brasilense Sp.KUL 9.

Table 1. Cell volumes of Azospirillum strains on various media as measured by the Coulter Counter Channelyzer

Species and Strain	Medium	Cell volume μm^3	relative volume value
Azospirillum lipoferum			
strain Sp.59b	Plate-Count-agar	0.79	1.0
Azospirillum brasilense			
strain Sp.KUL 9	Plate-Count-agar	1.58	2.0
strain Sp.KUL 9	Peptone-based	2.24	2.8
strain Sp.KUL 9	Beef-extract-based	2.37	3.0

Mean Cell Volume and MCC = Mean Cell Channel) obtained in some experiments are given. As evident from the table, the *A. brasilense* Sp.KUL 9 cells grown on PCA are about twice as big in volume as compared with those of *A. lipoferum* Sp59b grown on the same medium. Moreover, cells of Sp.KUL 9 grown on beef-extract agar have a volume of about 1.5 times greater than that of the same strain on PCA. The latter cells have now a volume 3 times larger than those of Sp. 59b on PCA. The generation times and the doublings per hour of both strains grown on some of these complex media is presented in Table 2. As evident from the table, the generation time of *A. brasilense* Sp.KUL 9 on peptone-based medium is greater than that of the same strain on other media. This increase of generation time is concomitant with an increase in cell size.

Discussion and taxonomic implications

As evident azospirilla are very pleomorphic organisms. Besides pleomorphism, some strains are in cell morphology phenotypically very different from other congeners. For instance, A. lipoferum Sp.41 produces rather large cells, whereas the cells of A. lipoferum Sp.59b are usually rather small. Moreover, A. lipoferum Sp B7c produces in most media rather round, coccoid cells resembling those of representatives of the Azotobacteriaceae. A further analysis of the latter strain is required.

In complex media the cells of Azospirillum (especially those of A. lipoferum) become larger and pleomorphic. This is also a characteristic of many Azotobacteriaceae grown under similar conditions. By means of cell volume measurements and concomitant growth rate estimates, we observed that the larger, pleomorphic cells required a larger generation time as compared to smaller cells of the same species grown on other media. Schaechter et al. (21) and Kjeldgaard et al. (15) studied cell mass, the average number of nuclei/cell and the content of RNA and DNA in Salmonella cells in relation to the composition of the medium and temperature. They confirmed the well-known observation that 'exponentially growing cells' are bigger in size than 'resting cells'. Studying the effect of the medium under unrestricted growth conditions and in restricted, balanced growth, they found the rate of protein synthesis to be directly proportional to the amount of RNA, whereas the DNA level is maintained for well-defined periods. In other words, the cells must grow bigger and increase their RNA/nucleus ratio, in order to grow faster. Superficially, it seems a paradox that the fewer the synthetic activities required during growth the bigger the cell. Our observations with cell size (volume) and growth of Azospirillum on complex media confirm this general phenomenon. The bigger, pleomorphic cells showed an increased generation time, i.e. a decrease in number of doublings per hour (Table 2). It may, however, be long before the mechanisms invoked by the medium to account for such transition pattern of growth can be discussed in specific chemical terms.

Tarrand et al. (22) assigned the genus name Azospirillum for Beijerinck's Spirillum lipoferum and allocated this genus in the section 'Aerobic/microaerophilic, motile, helical/vibrioid Gram-negative bacteria' (16). They consider it closely related to representatives of the genera Aquaspirillum, Spirillum, Oceanospirillum, Campylobacter, Bdellovibrio, and Vampirovibrio. Ribosomal RNA/DNA hybridization studies by De Smedt et al. (9) indicated that azospirilla belong to the same rRNA superfamily as Beijerinckia, Rhizobium, Agrobacterium, Acetobacter, Gluconobacter,

Table 2. Generation time and doublings/hour of two <u>Azospirillum</u> strains on various media at 34°C on basis of extinction and cell count

Strain	Medium	Extinction OD_{660}		Cell count	
		Generation time (hours)	Doublings/ hour	Generation time (hours)	Doublings/ hour
<u>Azospirillum</u>	YEGL	1.7	0.59	0.9	1.11
<u>brasilense</u>	YEMA	1.6	0.63	0.6	1.67
Sp. KUL 9	Peptone	2.4	0.42	1.0	1.00
	Beef-extr.	2.0	0.50	1.3	0.77
	N-free medium	c. 7.0 days	c. 0.006	c. 0.5 day	c. 0.08
<u>Azospirillum</u>	YEGL	1.2	0.83	0.9	
<u>lipoferum</u>	YEMA	1.7	0.59	1.0	1.00
Sp. 59b	Peptone	2.0	0.50	1.8	0.56
	Beef-extr.	1.5	0.67	1.9	0.53
	N-free medium	c. 3.5 days	c. 0.01	c. 1.0 day	c. 0.04

Note: YEGL = Yeast-extract glucose medium; YEMA = Yeast-extract malate medium.

<u>Xanthobacter</u>, and <u>Aquaspirillum</u>. Krieg and Döbereiner (16) suggested that members of the latter genus might be the closest-known relatives of <u>Azospirillum</u>. The latter hypothesis is supported by the observation that there is a relationship between <u>Azospirillum</u> and <u>Rhodospirillum</u> as indicated by oligonucleotide catalogs from 16 S rRNA, moreover by phenotypic resemblance, and by the striking similarity concerning the activation factor for the for the Fe protein of nitrogenase. However, it should be noted that studies on complementarity between Mo-Fe protein and the Fe protein of N_2-fixing organisms have emphasized the close similarity of proteins among

organisms of diverse origins. Cross reactions between these nitrogenase components are effective within the Azotobacteriaceae, but also with *Klebsiella pneumoniae*, *Bacillus polymyxa*, *Clostridium pasteurianum*, *Rhodospirillum rubrum*, and *Xanthobacter flavus* (formerly *Mycobacterium flavum*).

Besides these supposed similarities, there are also striking dissimilarities. For instance, in *Spirillum* the mol.% G + C of the DNA is c.38, in *Aquaspirillum* 49-66, in *Rhodospirillum* c. 62-66, and in *Oceanospirillum* 42-51, whereas in *Azospirillum* 69-71. Phenotypic characteristics of azospirilla with respect to other taxonomic groups seem to be of little value, because of the pleomorphism of *Azospirillum* and the occurrence of similar forms in genera of spiril-like organisms, vibrios, *Xanthobacter* and many other species. Diazotrophic abilities seem also not to be a conclusive criterion, because this property is also found in many unrelated organisms such as *Aquaspirillum*, *Rhodospirillum*, *Xanthobacter*, *Derxia*, *Azotobacter*, *Azomonas*, *Beijerinckia*, *Rhizobium*, some methanogenic bacteria, and in *Frankia* species. With regard to the mol.% G + C of DNA in *Azospirillum* (69-71), the value observed in *Derxia* with 69-73 appears to be the closest to this taxon. Moreover, *Derxia* has to some extent the same pattern of pleomorphism as *Azospirillum* (3,5). Further, the mol.% G + C of 65-70 in the genus *Xanthobacter* is close to the value found in *Azospirillum*. With *Azospirillum*, members of the genus *Xanthobacter* show vibrioid forms and the ability to produce 'snapping' cell configurations (see for *Azospirillum* this treatise). Besides this, similar to this genus certain *Azospirillum* strains are able to grow chemo-organotrophic as well chemo-lithoautotrophic utilizing the H_2-oxidation reaction (Knallgas reaction) as energy source and reducing power in CO_2 fixation (18, 19). But this ability is also present in various *Pseudomonas* and *Alcaligenes* species (formerly '*Hydrogenomonas*' species), nitrate-reducing *Paracoccus* species, *Bradyrhizobium* (formerly *Rhizobium*) *japonicum*, and in *Derxia*. Non-glucose utilizing representatives of Azotobacteriaceae also exist. In fact it was the main characteristic of *Azomonas insignis* mentioned in the type description and it is one of the characters to differentiate it from other *Azomonas*/*Azotobacter* species (3). The species *Azomonas insignis* is very motile with one polar flagellum (or lophotrichous) and in addition it has a great preference for organic acids, especially malic acid. *Azotobacter paspali* has the same type of flagellation as Azospirillum. This organism is a well-known associate of grass roots and shows also a great preference for organic acids. Finally, Franche and Elmerich (12) described glucose-utilizing, no biotin requiring Azospirillum strains, while another strain was observed to require other growth factors in addition of biotin. Wong et al. (25) isolated cellulolytic N_2-fixing Azospirillum strains and Nur et al. (20) obtained from grass roots vibrioid-shaped Azospirillum-like organisms having a much smaller (0.5 - 0.6 µm) cell diameter.

In conclusion, it can be stated that the taxonomy and the phylogenetic relationships of the taxon Azospirillum are not fully understood and rather unclear. On the basis of some overall characteristics, a continuum of species seems to exist between Azospirillum, some spirilla, other oxidative or micro-aerophilic, Gram-negative, polar-flagellated bacteria such as aerobic pseudomonads, Xanthobacter, oxidative vibrios, and members of the nitrogen-fixing family of Azotobacteriaceae. Before we discriminate between fundamental and non-fundamental characters, the bounderies for the genera within these groups tend to become indistinct, when precise definitions are attempted.

Acknowledgements

I wish to thank Dr. L. Reynders (Univ. of Leuven, Belgium), Dr. T. Heulin (C.N.R.S., Nancy, France) and Prof. Dr. G. Jagnow (FAL, Braunschweig, B.R.D.) for sending me Azospirillum strains. I like to acknowledge the excellent assistance of Miss A.R.M.L. Hamers for the cell counts and the preparation of the growth curves.

References

1. Becking, J.H. (1962). Proc. 8th Int. Congr. Microbiol., Montreal, Abstract B 14.5.

2. Becking, J.H. (1963). Antonie v. Leeuwenhoek, J. Microbiol. Serology 29, 326.

3. Becking, J.H. (1981). In Starr, M.P., Stolp, H., Trüper, H.G., Balows, A. and Schlegel, H.G. (eds.): The Prokaryotes, A Handbook on Habitats, Isolation, and Identification of Bacteria, Springer-Verlag, Berlin, vol. 1, 795-817.

4. Becking, J.H. (1982). In Klingmüller, W. (ed.): Azospirillum I, Genetics, Physiology, Ecology, Birkhäuser, Basel, Experientia Suppl. 42, 130-149.

5. Becking, J.H. (1984). In Krieg, N.R. and Holt, J.G. (eds.): Bergey's Manual of Systematic Bacteriology, Williams and Wilkins, Baltimore, vol. 1, 311-325.

6. Beijerinck, M.W. (1922). Koninkl. Ned. Akad. Wetenschp. Verslag Gewone Vergadering Afdeling Nat. 30, 431-438.

7. Beijerinck, M.W. (1925). Zentralbl. Bakt. Parasitenk. Infektionskrankh., Jena, Abteilung II, 63, 353-359.

8. Berg, R.H., Vasil, V. and Vasil, I. (1980). Protoplasma 101, 143-163.

9. De Smedt, J., Bauwens, M., Tijtgat, R. and De Ley, J. (1980). Int. J. Syst. Bacteriol. 30, 106-122.

10. Döbereiner, J. (1983). In Klingmüller, W. (ed.): Azospirillum II, Genetics, Physiology, Ecology, Birkhäuser, Basel, Experientia Suppl. 48, 9-23.

11. Eskew, D.L., Focht, D.D. and Ting, I.P. (1977). Appl. Environmental Microbiol. 34, 582-585.

12. Franche, C. and Elmerich, C. (1981). Ann. Microbiol. (Inst. Pasteur) 132, 3-18.

13. Heulin, T. (1983). Etude de quelques *Azospirillum* et de leurs effets spécifiques sur l'exsudation racinaire du riz. Thesis, 117 pp.

14. Heulin, T., Weinhard, P. and Balandreau, J. (1983). In Klingmüller, W. (ed.): Azospirillum II, Genetics, Physiology, Ecology, Birkhäuser, Basel. Experientia Suppl. 48, 89-94.

15. Kjeldgaard, N.O., Maaløe, O. and Schaechter, M. (1958). J. gen. Microbiol. 19, 607-616.

16. Krieg, N.R. and Döbereiner, J. (1984). In Krieg, N.R. and Holt, J.G. (eds.): Bergey's Manual of Systematic Bacteriology, Williams and Wilkins, Baltimore, vol. 1, 94-104.

17. Lamm, R.B. and Neyra, C.A. (1981). Can. J. Microbiol. 27, 1320-1325.

18. Malik, K.A. and Schlegel, H.G. (1980). FEMS Microbiol. Lett. 8, 101-104.

19. Malik, K.A. and Schlegel, H.G. (1981). FEMS Microbiol. Lett. 11, 63-67.

20. Nur, I., Okon, Y. and Hennis Y. (1980). Can. J. Microbiol. 26, 714-718.

21. Schaechter, M., Maaløe, O. and Kjeldgaard, N.O. (1958). J. Gen. Microbiol. 19, 592-606.

22. Tarrand, J.J., Krieg, N.R. and Döbereiner, J. (1978). Can. J. Microbiol. 24, 967-980.

23. Tchan, Y-T and New, P.B. (1984). In Krieg, N.R. and Holt, J.G. (eds.). Bergey's Manual of Systematic Bacteriology, Williams and Wilkins, Baltimore, vol. 1, 220-234.

24. Thomas-Bauzon, D., Weinhard, P., Villecourt, P. and Balandreau, J. (1982). Can. J. Microbiol. 28, 922-928.

25. Wong, P.P., Stenberg, N.E. and Edgar, L. (1980). Can. J. Microbiol. 26, 291-296.

ERRATUM

In Azospirillum II, 1983, article Heulin et al., insert on page 91, after line 5, the following:

plates were incubated at 37°C for 72 hours under air. Carbohydrate acidification was determined to be positive when the medium turned yellow (pH < 5.0), weak when it became green and negative when it remained purple violet.

Under those conditions it appeared that strain 4B pHT2 was unable to acidify glucose, ribose and mannitol unlike the parental 4B strain (Table 4).

Table 4. Carbohydrate acidification by the 4B, 4B 40 N1 and 4B pHT2 strains of *A. lipoferum* (microtiter plates)

Strain \ Carbohydrate	Glucose	Fructose	Ribose	Arabinose	Mannitol	Sorbitol
4B	+	+	+	+	Weak	+
4B 40N1	+	+	+	+	Weak	+
4B pHT2	-	+	-	+	-	+

Optimum oxygen partial pressure for growth

In another experiment we studied the effect of oxygen partial pressure (pO_2) on growth of 4B and 4B pHT2; the bacteria were cultivated in a N-free liquid medium with glucose as the only C and energy source. Growth was estimated by bacterial protein measurements (5) and optical density at 570 nm after 24 h (3). The results (Fig. 1) show that 4B pHT2 was actually able to use glucose for growth but only at low partial pressures of O_2. Whereas the ability of 4B to grow on glucose increased from 0.5 to 8% pO_2, the growth of 4B pHT2 decreased and its best growth was observed at 1% initial partial pressure.

www.ingramcontent.com/pod-product-compliance
Lightning Source LLC
Chambersburg PA
CBHW061204240926
51846CB00006B/1107